WHERE DID THAT CHEMICAL GO?

WHERE DID THAT CHEMICAL GO?

A Practical Guide to Chemical Fate and Transport in the Environment

Ronald E. Ney, Jr., Ph.D.

Library of Congress Catalog Card Number 90-33433
ISBN 0-442-00457-5

Printed in the United States of America

Van Nostrand Reinhold
115 Fifth Avenue
New York, New York 10003

Van Nostrand Reinhold International Company Limited
11 New Fetter Lane
London EC4P 4EE, England

Van Nostrand Reinhold
480 La Trobe Street
Melbourne, Victoria 3000, Australia

Nelson Canada
1120 Birchmount Road
Scarborough, Ontario M1K 5G4, Canada

16 15 14 13 12 11 10 9 8 7 6 5 4 3 2 1

Library of Congress Cataloging-in-Publication Data

Ney, Ronald E., 1936–
Where did that chemical go? : a practical guide to chemical fate and transport in the environment / Ronald E. Ney, Jr.
p. cm.
Bibliography: p.
ISBN 0-442-00457-5
1. Pollution—Environmental aspects. I. Title.
TD174.N45 1990
628.5′2—dc20 90-33433
CIP

Contents

Preface

Chapter 1 demonstrates what the book is about—the use of techniques to predict the fate and transport of a chemical in the environment. Definitions are given in terms simple enough for any reader of the book to understand them. Chapters 2 and 3 present predictive techniques and ways of using them. Tables are given in which users can insert data to make the predictions. Chapter 4 ties together the material of the previous chapters, showing how data and predictive techniques may be combined to assess the route of exposure. Chapter 5 gives examples of specific chemicals, showing different types of data, and guides the user, as if hands on, through a step-by-step prediction procedure. The reader will learn how to use and interpret pertinent data and how to predict the fate and transport of a chemical, as well as to predict exposure routes.

The book is intended for use by either laypersons or scientists. It will enable Realtors®, real estate appraisers, lending institutes, developers, waste managements, chemical manufacturers and users, environmentalists, educators, government officials, environmental assessors, private citizens and others to predict potential routes of exposure to chemicals.

As you read the book, I hope you will enjoy answering the question, where did that chemical go?

Acknowledgments

I am grateful for the support of my wife, Sue McClure Ney, who provided not only her word processing skills but generous portions of encouragement, consideration, and love as well, during the preparation of this book. I also thank Rene Ramos for contributing the illustration of environmental compartments.

Chapter 1
Fundamentals

FATE AND TRANSPORT

Many questions must be considered when we attempt to determine what happens to a chemical and its breakdown products (i.e., degradates, biodegradates, metabolites, transformation products, dissociation products, hydrolytic products) in the environment. After we obtain data about such chemicals, it can be quite difficult to interpret them. The intention of this book is to present an overview of the data that are obtainable and of their interpretation.

We will: (1) discuss a chemical as the parent chemical because almost any organic chemical, parent or breakdown product, will fit the examples given; (2) present discussions on what could happen to a chemical in an environmental compartment; (3) show how to make predictions on chemical fate in the environment; and (4) demonstrate how, in most cases, to determine and predict the fate of an organic chemical in a particular environmental compartment when few or no data are available. The overall purpose of this book is that it should offer insight and understanding, as well as be a guide to predicting the fate and transport of chemicals (mostly organic chemicals) in the environment.

There are many questions to answer in the assessment of the fate of a chemical in environmental compartments. Six are considered here.

1. *What are the mechanisms that break down a chemical to a toxic, nontoxic, or naturally occurring chemical(s)* [1]? These mechanisms include the following:

a. Photolysis. If a chemical can sorb sunlight, the chemical can be transformed in media (i.e., air, water, and soil) or on media

(i.e., water, soil, plants, and animals) to produce transformation products.
b. Biodegradation in all environmental compartments to produce biodegradates, if microbes are present.
c. Metabolism in plants and animals to produce metabolites.
d. Hydrolysis in all environmental compartments to produce hydrolytic products.
e. Dissociation in all environmental compartments to produce dissociated products.
f. Sorption in all environmental compartments.
g. Bioaccumulation in plants and animals.

2. *What is the rate of dissipation or half-life* ($t_{1/2}$) *of a chemical in environmental compartments* [1]? The rate of dissipation or the half-life is the time that it takes for the chemical to be reduced by one-half of its original amount, measured from the time of its introduction into an environmental compartment (i.e., when 50%) of the chemical has vanished and only 50% of it remains. The lost 50% could be in the form of a breakdown product(s), either toxic, nontoxic, or naturally occurring. This is why in real situations breakdown products must be identified and studied in the same way that the parent chemical is studied. The dissipation of a chemical may be real, resulting in the breakdown of the parent chemical (usually organic chemicals), or not real, due to adsorption or movement of the chemical from contaminated areas to noncontaminated areas. There are ways to predict this, even when no or few data are available, as will be discussed later. In any event, these aspects should be determined by sorption or mobility studies. One must always remember that the fact that a chemical cannot be found does not mean it is not there. Good testing will establish its whereabouts.

3. *How is a chemical moved in the environment by natural means* [1]? These means include:

a. Volatilization: A chemical that is volatile can move into the air; and once in the air, the chemical can be transformed, hydrolyzed, dissociated, biodegraded, or sorbed to dust particles. The chemical also can be released from air to contaminate soil, water, plants, and animals via dust particle fallout or by precipitation.

b. Leaching: A chemical may be transported through soil by solvents (water, etc.) or with soil movement. Any chemical that leaches can contaminate groundwater, an action that in turn may result in the contamination of air, soil, other waters, plants, and animals.
c. Runoff: A chemical may move across a surface with a solvent or with soil movement to contaminate air, water, soil, plants, and animals.
d. Food-chain contamination: A chemical that is passed from one environment can be passed on to water, air, soil, plants, or animals and on up the food chain. For example:
   (1) A chemical that gets into the aquatic environment can be taken up by microscopic organisms, which are eaten by fish, which are eaten by wildlife and humans.
   (2) A chemical in water could reach crops via irrigation, and the crops could be eaten by wildlife and humans.

4. *What happens to a chemical adsorbed or absorbed in soil or in plants or animals* [1]? We will only discuss sorption in soil; because sorption in plants and animals is beyond the scope of this book although it will be discussed in general, in terms of bioaccumulation. These are the possibilities in soil:

a. Absorption: A chemical that is absorbed in soil can be released in any environment (like water being released from a sponge).
b. Adsorption: A chemical that is adsorbed to soil particles is unlikely to be leached by water. An adsorbed chemical may be leached with other solvents (if they get into soil), by soil movement, or by water if soil sorption sites have been filled or taken up by another chemical. Another cause of leaching is saturation; that is, leaching occurs as the starting chemical is continuously added to the soil, thus saturating the soil sites. (Consider the analogy of a parking lot that is full; where do the cars park?) A plant can release soil-bound residues via its root system and translocate the chemical throughout the plant. Likewise, animals that ingest soil can release the chemical residues adsorbed in the soil and translocate them throughout their bodies. These residues could be absorbed in the fat tissue or by protein in the animals [2].

5. *How does food-chain contamination occur* [1]? This type of contamination occurs when one environment is contaminated by a chemical, and that chemical is released into another environment, the end result being that animals ingest the chemical. For example:

a. A chemical that gets into air can be released onto soil, water, plants, and animals, and then noncontaminated animals may eat, drink, or breathe the contamination.
b. Fish may bioaccumulate the chemical in their tissue, and when other animals eat the fish, they in turn also will become contaminated. Thus food-chain contamination has occurred, followed by bioaccumulation up the food-chain ladder.

6. *What are accumulation and bioaccumulation?* These terms refer to the condition in which a chemical builds up and does not dissipate; in soil it is called accumulation, and in plants or animals it is called bioaccumulation. The chemical is not broken down in an environmental compartment. In mammals another problem exists—the potential for residues to be passed on into mothers' milk, then to nursing infants.

REFERENCES

[1] Ney, Ronald E. Jr., "Regulatory Aspects of Bound Residues," Workshop V-C, 4th International Congress of Pesticide Chemistry, International Union of Pure and Applied Chemistry, Zurich, Switzerland, Unpublished paper, 1978.
[2] Yip, George and Ronald E. Ney, Jr., "Analysis of 2,4-D Residues in Milk and Forage," *Weeds, Journal of the Weed Society of America,* vol. 14, pp. 167–170, April 1966.

ENVIRONMENTAL COMPARTMENTS

Five environmental compartments are considered herein. Many chemical, physical, and living reactions can occur in these compartments; and before any data can be discussed, we must know what the compartments are, and what could occur in each one. We now consider relevant features of each compartment.

Compartment Air

- *Contamination:* Air contamination occurs when a volatile chemical or airborne particulate matter (dust) containing a chemical gets into the air as a result of a spill, evaporation, or any release.
- *Reactions:* A chemical can be phototransformed in air, or it can be sorbed to particulate matter and be biodegraded, dissociated, hydrolyzed, or phototransformed.
- *Mobility:* Chemicals can be moved throughout air, by air or precipitation, or can move as fallout with precipitation or with particulate matter to contaminate other environmental compartments.
- *Exposure:* Airborne chemicals could result in chemical exposure of all environmental compartments [1].

Compartment Water

- *Contamination:* Water contamination occurs by fallout from air, from spills, from substances directly applied or intentionally put into water, with runoff, or from leaching into water.
- *Reactions:* Depending on the water (i.e., streams, lakes, ponds, groundwater, ocean, etc.), reactions may include dissociation, hydrolysis, phototransformation, biodegradation, or sorption to particulate matter.
- *Mobility:* Movement may occur with volatilization, water movement, evaporation, irrigation with well water, or animals, resulting in environmental contamination.
- *Exposure:* Chemicals in this compartment could result in the chemical exposure of all environmental compartments [1].

Compartment Soil

- *Contamination:* Soil contamination occurs by spills, fallout from air, or substances directly or indirectly applied to or put into or on soils.
- *Reactions:* Hydrolysis, dissociation, sorption, biodegradation, and photolysis reactions are possible.
- *Mobility:* Volatilization, runoff, leaching, or plant and animal uptake resulting in food-chain contamination may occur.

- *Exposure:* Soil contamination could result in the chemical exposure of all environmental compartments [1].

Compartment Plants

- *Contamination:* Plant contamination may result from fallout, spills, substances indirectly or directly applied to soils, irrigation, and materials in manure and in compost.

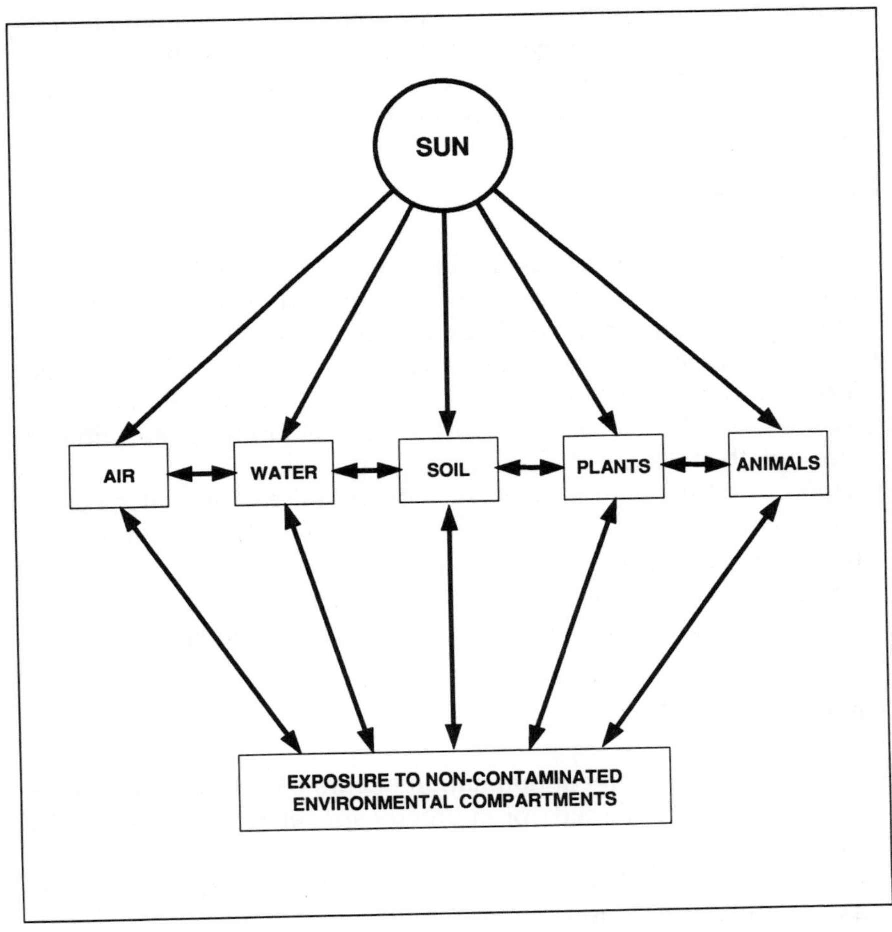

Figure 1-1. Environmental compartments. A never ending circle, from self-contamination to contamination of noncontaminated compartments and back again.

- *Reactions:* Metabolism, hydrolysis, dissociation, photolysis on the plant surface, or sorption may occur.
- *Mobility:* Movement may be occur by release into the air, into the soil via the root system, into the air if the plant is burned, or into the food-chain if the plant is eaten.
- *Exposure:* Plant contamination could result in the chemical exposure of all environments [1].

Compartment Animals

- *Contamination:* Animal contamination may occur by fallout, substances directly or indirectly applied to or on animals, eating other plants or animals, drinking water, or breathing.
- *Reactions:* Metabolism, hydrolysis, dissociation, bioaccumulation, and phototransformation on surfaces are possible.
- *Mobility:* Respiration, excretion, and release if the substance is burned or eaten are means of movement.
- *Exposure:* Animal contamination could result in exposure of all environmental compartments [1].

Thus we have seen: how environmental compartments are contaminated; reactions that could occur to chemicals in those compartments; mobility in and out of the compartments, and how one compartment can contaminate other noncontaminated compartments; and the exposure possibilities (see Figure 1-1 and Table 1-1).

Table 1-1. Reactions in Environmental Compartments.

Reactions	Air	Water	Soil	Plants	Animals
Hydrolysis	X	X	X	X	X
Phototransformation	X	X	X	X	X
Dissociation	X	X	X	X	X
Solubility	X	X	X	X	X
Sorption	X	X	X	X	X
Biodegradation	X	X	X	X	X
Metabolism				X	X
Accumulation			X		
Bioaccumulation				X	X
Volatilization	X	X	X		
Respiration				X	X
Excretion					X

REFERENCES

[1] Ney, Ronald E., Jr., "Exposure Assessment Considerations and Problems," Exposure Assessment Workshop, U.S. Environmental Protection Agency, UP, 1982.

Chapter 2
Physical and Chemical Processes

WATER SOLUBILITY

Water solubility is perhaps the most important chemical characteristic, used to assess (1) chemical mobility, (2) chemical stability or breakdown, (3) chemical accumulation, (4) chemical bioaccumulation, and (5) chemical sorption in any environmental compartment. Water solubility should be one of the easiest chemical test methods; however, if you were to look up the solubility of DDT, you would find a long list of solubilities. Luckily those values are presented in a range that enables one to make some kind of assessment or prediction; and that is the purpose of our discussions.

What can water solubility be used for? Remember these key points:

1. The higher the water solubility of a chemical, the more likely it is to be mobile, and the less likely it is to be accumulative, bioaccumulative, volatile, and persistent; and a highly soluble chemical is prone to biodegradation and metabolism that may detoxify the parent chemical.
2. The lower the water solubility of a chemical, the more likely it is that it will be immobilized via adsorption, and thus it is less mobile, more accumulative or bioaccumulative, persistent in environmental compartments, and slightly prone to biodegradation, and it may be metabolized in plants and animals.
3. The in-between range of high and low water solubilities indicates chemicals whose behavior could go either way, as discussed in 1 and 2 above.

The above generalities are meaningless unless we have values or ranges on which to base predictions. Thus, to discern the fate of chemicals by water solubilities, let us assign the following numerical values, presented by Ney [3]:

1. Low solubility: less than 10 ppm (<10 ppm).
2. Medium solubility: between 10 and 1,000 ppm.
3. High solubility: greater than 1,000 ppm (>1,000 ppm).

Table 2-1 may be used to link the environmental fate of chemicals to their water solubility (WS).

The relationships are not clear-cut, as the water solubility of a chemical can also affect other degradative or transformation process, such as hydrolysis, photolysis, and/or biodegradation. Water solubility can be used to predict sorption or desorption with soil organic matter/organic carbon, mobility/leaching, and bioaccumulation in animals by means of lipo-solubility prediction. These aspects of its use will be discussed in other chapters.

Proven, validated test methods should be used to measure the water solubility of a chemical. If at all possible, a radiolabeled parent chemical should be used. The use of a radiolabeled chemical will enable the researcher to discern whether sorption to equipment, phototransformation (easily prevented), or dissociation has occurred, and, most important, to validate chemical stability, solubility, and analytical methodology. A radiolabeled chemical should be used to validate most studies.

Table 2-1. Water Solubility (WS)

ENVIRONMENTAL COMPARTMENTS	LOW <10PPM	MEDIUM 10–1,000PPM	HIGH >1,000PPM
Mobility	n*	either way	yes
Adsorption	yes	either way	n
Biodegradation	maybe	either way	yes
Metabolism	maybe	either way	yes
Accumulation	yes	either way	n
Bioaccumulation	yes	either way	n
Persistence	yes	either way	n
Dissipation	n to slowly	either way	yes
Food-chain contamination	yes	either way	n

* n denotes negligible.

In the environment there would be many effects on chemical solubility, such as the effects of temperature, the concentration of chemicals, sorption, and so on.

If no data on solubility exist, then one could predict solubility by using chemical structure methodology, an advanced predictive technique for chemists. I have found the best illustrations in the *Handbook of Chemical Property Estimation Methods* by Warren J. Lyman, William F. Reehl, and David H. Rosenblatt, published by the McGraw-Hill Book Company, a source that will be cited frequently in this book [2]. Kenaga [1] also has reported on mathematical equations used to calculate solubility.

Lyman et al. [2] reported dichloro diphenyl trichloroethane (DDT) to have a measured solubility of 1.2 μg/L or 1.7 μg/L, indicating a large range for error (200%). Lyman et al. [2] also reported a water solubility, using the chemical structure of DDT, of 3.89 μg/L. Other sources report that DDT is almost insoluble. For the purpose of this discussion, let us use the solubilities of 1.2 μg/L and 3.89 μg/L to predict what may happen in the environment. To do this, we will convert μg/L to ppm, for use with Table 2-1. The following example will help you to understand how to convert μg/L (ppb) to mg/L (ppm) (in this case, 1.2 μg/L):

ppm = mg/L (parts per million = milligrams per liter)
ppb = μg/L (parts per billion = micrograms per liter)
μg/L = one millionth of a gram per liter
mg/L = one thousandth of a gram per liter

$$\frac{1{,}000 \text{ mg/L}}{1{,}000{,}000\ \mu\text{g/L}} \times 1.2\ \mu\text{g/L} =$$

$$\frac{0.001 \text{ mg/L} \times 1.2\ \mu\text{g/L}}{\mu\text{g/L}} = 0.0012 \text{ mg/L (ppm)}$$

DDT has a water solubility in the range of 0.0012 ppm to 0.00389 ppm; thus, it falls at the level of <10 ppm (see Table 2-1).

With this one piece of data, we can say that DDT will persist in any environmental compartment, accumulate in soil, bioaccumulate in plants and animals, and bioconcentrate in the food-chain. The same data can be used in other predictive methods, which will be discussed in following chapters.

If the solubility of a chemical were 1,000 ppm, then predictions about its fate could be based on Table 2-1. If it were in the range of 10 to 1,000 ppm, problems could arise, and additional data might be needed to predict its fate in environmental compartments. The use of other predictive methods will be discussed in following chapters.

REFERENCES

[1] Kenaga, E. E. and C. A. I. Goring, "Relationship Between Water Solubility, Soil Sorption, Octanol–Water Partitioning, and Concentration of Chemicals in Biota," Special Technical Publication 707, American Society for Testing and Materials, 78–115, 1980.

[2] Lyman, Warren J., William F. Reehl, and David H. Rosenblatt, *Handbook of Chemical Property Estimation Methods,* McGraw-Hill Book Company, 2–44 and 2–45, 1982.

[3] New, Ronald E., Jr., "Fate, Transport and Prediction Model Application to Environmental Pollutants," Spring Research Symposium, James Madison University, Harrisonburg, Virginia, UP, April 16, 1981.

OCTANOL WATER

The octanol water partition coefficient (Kow), or partition coefficient (P), is an indicator of the bioaccumulation or bioconcentration potential of a chemical in the fatty tissue of living organisms. The Kow or P value (which has no units) is an indicator of water solubility, mobility, sorption, and bioaccumulation. The symbols Kow and P are used interchangeably, but this book will use only Kow.

The Kow represents a mathematical equation expressing the ratio of the equilibrium concentrations of a chemical in octanol and water phases; that is, it is the ratio of the concentration of a chemical in octanol to the concentration of that chemical in water at equilibrium. Simply put, it is the ratio of an organic chemical's distribution between octanol and water phases:

$$\text{Kow} = \frac{\text{concentration of organic chemical in octanol phase}}{\text{concentration of organic chemical in water phase}}$$

The use of the Kow depends on its size:

1. The higher the Kow is, the greater the affinity of the chemical to bioaccumulate/bioconcentrate in the food chain, the greater its potential for sorption in soil, and the lower its mobility. This also means lower solubility in water. Do not confuse water solubility with Kow, as the ratio does not express water solubility, and there are no units of measure, just a number.
2. The lower the Kow is the less the chemical's affinity to bioaccumulate, the greater its potential for mobility, the greater its solubility, and the greater its potential to biodegrade and to be metabolized by plants and animals.

Here also we need numbers for assessment of chemical fate; thus, I have made the following numerical assignments [3]:

1. A Kow of less than 500 (<500) would be indicative of high water solubility, mobility, little to no bioaccumulation or accumulation, and degradability by microbes, plants, and animals.
2. A high Kow, greater than 1,000 (>1,000), is indicative of low water solubility, immobility, nonbiodegradability, and a chemical that is bioaccumulative, accumulative, persistent, and sorbed in soil.
3. A midrange Kow, 500 to 1,000, indicates that the chemical can go the way of either low or high Kow.

Table 2-2 show how to predict the environmental fate of chemicals by using the Kow.

Table 2-2. Octanol Water Partition Coefficient and Fate of Chemicals

ENVIRONMENTAL COMPARTMENTS	LOW KOW <500	MEDIUM KOW 500–1,000	HIGH KOW >1,000
Persistent	n*	either way	yes
Adsorbed	n	either way	yes
Absorbed	yes	either way	n
Biodegraded	yes	either way	n to slowly
Metabolized	yes	either way	n to slowly
Accumulated	n	either way	yes
Bioaccumulated	n	either way	yes
Dissipated	yes	either way	n

* n denotes negligible.

There are many ways to obtain a Kow, by either laboratory test or mathematical equations. Proven and validated test methods should be chosen. If at all possible, a radiolabeled parent chemical should be used; the reasons for doing so were discussed in the section on water solubility.

If no data exist, chemical structure can be used to predict the Kow, [2]. Lyman et al. [2] and Kenaga [1] have reported on the use of mathematical equations for Kow prediction.

Kenaga [1] reported DDT to have a Kow of 960,000. Using Table 2-2, one can easily see that a Kow of 960,000 is >1,000, indicating that DDT has low water solubility, is persistent, is adsorbed in soil, is slowly biodegraded, is slowly metabolized, and is accumulated and bioaccumulated, so that food-chain contamination can be predicted. These are a different expression of the predictions based on the water solubility of DDT.

REFERENCES

[1] Kenaga, E. E. and C. A. I. Goring, "Relationship between Water Solubility, Soil Sorption, Octanol–Water Partitioning, and Concentration of Chemicals in Biota," Special Technical Publication 707, American Society for Testing and Materials, 78–115, 1980.

[2] Lyman, Warren J., William F. Reehl, and David H. Rosenblatt, *Handbook of Chemical Property Estimation Methods,* McGraw Hill Book Company, Chapter 1, 1982.

[3] Ney, Ronald E., Jr., "Fate, Transport and Prediction Model Application to Environmental Pollutants," Spring Research Symposium, James Madison University, UP, April 16, 1981.

HYDROLYSIS

Hydrolysis is perhaps one of the most important mechanisms in the environment for the breakdown of a parent chemical [2]. It occurs in soils, water, plants, animals, and possibly air because water exists in all of these environments. The hydrolysis of pesticides has even occurred on plant surfaces.

Many environmental factors influence the rate of hydrolytic degradation, such as temperature, pH, solubility, sunlight, ad- or absorption, volatility, and so on. The rate of hydrolization of a chemical is the time that it takes to reach one-half ($t_{1/2}$) of its original amount. Because many outside influences can affect the rate of hydrolysis, study of the process should be done under controlled conditions using a radiolabeled parent molecule. The use, if applicable, of a radiolabeled chemical enables the researcher to study sorption, breakdown products, and recovery (to validate analytical methods).

There are chemical structure prediction methods and mathematical equations available to predict hydrolysis, as reported by Lyman et al. [1], but these approaches are for the experienced professional.

The rate of hydrolysis must be known to determine persistence in the environment. (Persistence is how long a parent chemical will be present for exposure.) There are no clear cut-assessments for half-lives because exposures differ in the environmental compartments. The rate of hydrolysis can be used in exposure assessments for toxic and phytotoxic chemicals, but toxicity considerations are not discussed herein, as they are beyond the scope of this book.

Water Compartments

No half-life is acceptable if the chemical causes immediate harm to organic organisms, or harm if immediately used for irrigation or potable water. If there are no immediate concerns, then a half-life of 30 days may be acceptable if no accumulation/bioaccumulation occurs. Because water cannot be controlled, care has to be taken in assessing exposure. It would be best to keep contaminating chemicals out of water.

Soil Compartments

In most cases soils can be controlled; thus the half-life of a parent chemical in soil can be considered differently from that in water or air. To control soil also means to control edible crops and animals in the area. If the chemical is immediately taken up and causes harm to plants or animals, then no half-life is acceptable. If a chemical is immediately taken up by plants and animals and metabolized without harmful breakdown products, and with no harm to plants and animals, then its presence may be acceptable. If the chemical has a half-life greater than 60 days, then the future use of the land must be considered for a chemical taken up by plants and animals. If the chemical breaks down in 1 hour, 1 day, 100 days, or longer and causes no environmental harm, then the half-life may not make a difference.

Air Compartment

This compartment is not usually studied; however, if a chemical is immediately harmful to plants and animals, no half-life is acceptable. Also, if it is not harmful but results in food-chain contamination, then no half-life is acceptable. Air cannot be controlled; so it is best to keep contaminating chemicals out of this compartment.

Plant and Animal Compartments

These compartments usually are studied for metabolism if the chemical is a pesticide. No discussions are presented herein on metabolism, as that subject is beyond the scope of this book.

The faster the hydrolytic rate is, the less likely the possibility of continued exposure in the enviornment. If the hydrolytic half-life ($t_{1/2}$ is <30 days, then accumulation, bioaccumulation, and food-chain contamination are not likely; if $t_{1/2}$ is between 30 and 90 days, chemical behavior goes either way; and if $t_{1/2}$ is $>$ days, contamination is likely. Table 2-3 will help one to visualize this assessment.

Table 2-3. Hydrolytic Half-Life (Days) and Fate of Chemicals.

ENVIRONMENTAL COMPARTMENTS	RAPID $t_{1/2} < 30$	MEDIUM $t_{1/2}$ 30 TO 90	SLOW $t_{1/2} > 90$
Accumulation	not likely	either way	yes
Bioaccumulation	not likely	either way	yes
Food-chain contamination	not likely	either way	yes
Persistence	n*	either way	yes
Adsorption	n	either way	maybe
Dissipation	yes	either way	n to slowly
Problem if acutely hazardous	yes	either way	yes
Problem only if hazardous via long term	not likely	maybe	likely

* n denotes negligible.

REFERENCES

[1] Lyman, Warren J., William F. Reehl, and David H. Rosenblatt, *Handbook of Chemical Property Estimation Methods,* McGraw Hill Book Company, Chapter 7, 1982.

[2] Ney, Ronald E., Jr., "Fate, Transport and Prediction Model Application to Environmental Pollutants," Spring Research Symposium, James Madison University, UP, April 16, 1981.

PHOTOLYSIS

A chemical can be phototransformed as long as it can absorb sunlight. Phototransformation of a chemical can occur in air, soil, or water and on surfaces of water, soil, plants, and animals. The photolytic product(s) can be either a higher- or a lower-molecular-weight chemical(s). This process also has been called photodegradation. Photooxidation can occur in the stratosphere, but is not considered in this discussion.

A chemical has to sorb light to be phototransformed; thus, if a chemical can be analyzed spectrophotometrically, it has a good chance to be phototransformed. Environmental influences can have an effect on the rate of phototransformation, such as depth of the chemical in soil and in water, sorption to soil, sensitizers, quenchers, and pH. The rate of phototransformation is the time that it takes for a parent chemical to be transformed to one-half (the half-life, $t_{1/2}$) of its original amount. This rate could differ in soil, in water, and on surfaces. The rate of photolysis can be used to determine persistence in the environment.

The considerations used for hydrolysis also apply to photolysis. The faster the photolytic rate is, the less likelihood there is of continued exposure in the environment. If the photolytic half-life is <30 days, then accumulation, bioaccumulation, or food-chain contamination is unlikely; if it is 30 to 90 days, the chemical behavior goes either way; and if it is >90 days, contamination is likely. Table 2-4 will help you to visualize the possibilities.

For a technical review of the photo process, Lyman et al. [1] go into explicit detail.

Table 2-4. Photolytic Half-Life (Days) and Fate of Chemicals.

ENVIRONMENTAL SURFACES	RAPID $t_{1/2} < 30$	MEDIUM $t_{1/2}$ 30 TO 90	SLOW $t_{1/2} > 90$
Accumulation	not likely	either way	yes
Bioaccumulation	not likely	either way	yes
Food-chain contamination	not likely	either way	yes
Persistence	n*	either way	yes
Adsorption	n	either way	maybe
Dissipation	yes	either way	slowly
Problem if acutely hazardous	yes	either way	yes
Problem only if hazardous via long term	not likely	maybe	likely

* n denotes negligible.

REFERENCE

[1] Lyman, Warren J., William F. Reehl, and David H. Rosenblatt, *Handbook of Chemical Property Estimation Methods,* McGraw Hill Book Company, Chapter 8, 1982.

VOLATILIZATION

A chemical on or in soil, on or in water, or on plants or animals may volatilize and get into the air. Vapor pressure is one of the most important factors governing volatilization, and provides an indication of whether a chemical will volatilize into the air under environmental conditions.

Some of the factors that effect volatilization in the environment are climate, sorption, hydrolysis, and phototransformation [1]. Here are guidelines:

1. A chemical with a low vapor pressure (VP), high adsorptive capacity, or high water solubility is less likely to volatilize into the air.
2. A chemical with a high VP, low sorptive capacity, or very low water solubility is more likely to volatilize into the air.
3. Chemicals that are gases at ambient temperatures will get into the air. (Gases at ambient temperature are not considered herein.)

Rapid volatilization into the air could result in immediate hazards for chemicals released indoors or outdoors if workers are in the area (e.g., agricultural workers). Vapor pressure is reported in terms of mm Hg (millimeter or mercury) or torr (which is equivalent to mm Hg). Table 2-5 will help you to visualize the possibilities.

REFERENCE

[1] Ney, Ronald E., Jr., "Fate, Transport and Prediction Model Application to Environmental Pollutants," Spring Research Symposium, James Madison University, UP, April 16, 1981.

SOIL SORPTION

The physical-chemical process by which a soil(s) ties up chemicals so that they are not released or are very slowly released in the environment is called adsorption, or bound residues. A chemical that is held by soil but is easily released is absorbed (in the way that a sponge holds water). The release mechanism considered herein is water, which is available in nature.

Plant root systems, as reported by Ney [4], have been shown to release adsorbed chemicals from soil for uptake into plant parts.

Table 2-5. Vapor Pressure (mm Hg or torr) and Fate of Chemicals.

ENVIRONMENTAL SURFACES	LOW < 0.000001	MEDIUM 0.000001 TO 0.01	HIGH > 0.01
Volatility	low	medium	high
Accumulation	yes	maybe	n*
Bioaccumulation	yes	maybe	n
Food-chain contamination	yes	maybe	n
Persistence	yes	maybe	n
Adsorption	high	maybe	low
Dissipation	yes	maybe	n
Problem if acutely hazardous	n	maybe	yes
Problem only if hazardous via long term	yes	maybe	n
Solubility	high	medium	low

* n denotes negligible.

Animals ingesting soil have also released the adsorbed chemical(s) for uptake. The mobility of adsorbed chemicals is prevented in soils. Here, the use of the word soil means only those soils that can adsorb chemicals, because soils do not all adsorb chemicals (e.g., sand, low-organic soils, etc., do not).

The movement of soil or soil particles containing an adsorbed chemical can contaminate other environments (e.g., as soil runoff or airborne particulates).

Sorption to soil, in almost all cases, can prevent phototransformation, hydrolysis, volatilization, mobility by water solubility, and microbial biodegradation.

Prior to the 1970s, many scientists would report that soils studied contained no chemical residues when, in fact, the scientists did not consider adsorbed residues or breakdown products. As previously discussed, the only way to discern what really occurs is to use a radiolabeled parent chemical. If the chemical cannot be extracted from soil with water or with other solvents, then there are two ways to discern the presence of a bound residue: (1) by combustion of the soil to measure adsorbed radioactivity, and (2) by planting crops to study any chemical residue(s) taken up by the crops (only needed if crops are involved).

Water is used as an extraction solvent to study bound residues. It should be noted, however, that other solvents, which do not occur naturally in the environment, may solubilize the adsorbed chemical for extraction. In most cases such solubilization will not happen, but it could be a problem in areas of solvent spills or in leaking storage area.

Soil sorption, chemical sorption, or bound chemical(s) in soil may be expressed as the extent that an organic chemical partitions between a solid phase and a liquid phase. This value is better known as the adsorption coefficient (*Koc*), and is expressed as the μg adsorbed per organic carbon (soil solid phase) divided by μg per ml of solution (liquid base).

Authors have variously reported absorption coefficients or adsorption to organic matter as *Kd, Koc,* or *Kom*. *Kom* is the μg adsorbed per organic matter (soil organic) divided by μg per ml of solvent (liquid). This book will use only *Koc*.

Many relevant mathematical equations are given by Lyman et al.

[2]. Helling and Turner [1] and Ney [3] have discussed soil sorption and leaching potential; Helling and Turner [1] have reported on soil thin layer chromatography (soil TLC). This is a very unique method, which could be used to combine many soil studies. Helling and Turner [1] divided soil mobility into classes, which will be discussed in a later chapter. These classes also can be used to predict adsorption. Comparing mathematical predictions, soil TLC, water solubility, and actual studies will help one to pinpoint erroneous scientific results.

We must remember that all soils do not act alike, nor do all chemicals. Thus, the chemical's identity and a study of soil characteristics are needed. Several soil characteristics may enhance sorption to a few chemicals, including the cation exchange capacity (CEC), clay-colloid-polymerization, clay with high surface areas, and so on.

To predict whether a chemical could be adsorbed to soil organic carbon (OC), one could use adsorption coefficients (Koc). The Koc value is only a numerical number with no units. Soils must contain organic carbon/organic matter to produce a Koc.

Chemical(s) with a high Koc of $> 10,000$ will adsorb to OC. Chemicals with a Koc in the range of 1,000 to 10,000 could behave either way. Chemicals with a low Koc of $< 1,000$ will not adsorb to soil OC. As discussed, these numbers for chemicals can be calculated or measured. When they are coupled with soil TLC and other studies, a counter-check system is formed.

Soil TLC gives a measure of the movement through soil, expressed as Rf or TLC-Rf with no units even though measured as millimeters or lesser units.

Chemicals with a high TLC-Rf, > 0.75, will not be adsorbed in soil and will be mobile (leach). Chemicals with a TLC-Rf in the range of 0.34 to 0.75 could go either way. Chemicals with a low TLC-Rf, < 0.34, will be adsorbed and should not leach with water.

Table 2-6 helps us to visualize what can be predicted with a Koc and a TLC-Rf.

Chemicals that can be desorbed from soil are chemicals that were not adsorbed. The release of chemicals by water is called desorption. These chemicals usually have a WS in the range of > 10 ppm and are mobile.

Table 2-6. Sorption to Soil and Fate of Chemicals.

SOILS AND SEDIMENTS	LOW Koc $> 10{,}000$ Rf < 0.34	MEDIUM Koc 1,000 TO 10,000 Rf 0.34 TO 0.75	HIGH Koc $< 1{,}000$ Rf $> .75$
Adsorption	yes	either way	n*
Mobility	n	either way	yes
Accumulation	yes	either way	n
Bioaccumulation	yes	either way	n
Food-chain contamination	yes	either way	n
Solubility	n	either way	yes
Persistence	yes	either way	n
Dissipation	n	either way	yes

* n denotes negligible.

REFERENCES

[1] Helling, Charles S. and Benjamin C. Turner, "Pesticide Mobility: Determination by Soil Thin-Layer Chromatography," *Science,* vol. 162, pp. 562–563, 1968.

[2] Lyman, Warren J., William F. Reehl, and David H. Rosenblatt, *Handbook of Chemical Property Estimation Methods,* McGraw-Hill Book Company, Chapter 4, 1982.

[3] Ney, Ronald E., Jr., "Fate, Transport and Prediction Model Application to Environmental Pollutants," Spring Research Symposium, James Madison University, UP, April 16, 1981.

[4] Ney, Ronald E. Jr., "Regulatory Aspects of Bound Residues," Workshop V-C, 4th International Congress of Pesticide Chemistry, International Union of Pure and Applied Chemistry, Zurich, Switzerland, UP, July 24–25, 1978.

LEACHING IN SOIL

The movement of a chemical downward through soil by water is called leaching. It is of concern because of the possibility that the chemical will move through the soil and contaminate the groundwater. If this happened, then well-water, aquatic-organism, and food-chain contamination could occur.

Many factors affect whether or not a chemical leaches in soil [1], including solubility, biodegradation, hydrolysis, dissociation, sorption, volatility, rainfall, and evapo-transpiration. Their effects on mobility are described as follows:

1. A chemical that is water-soluble can leach in soil and is likely to be biodegraded by soil microbes. If biodegradation is rapid, then leaching may not be a problem. Chemicals may be leached by solvents other than water, but other solvents are not considered herein as leaching solvents.
2. A chemical that is insoluble in water can be adsorbed in soil, moved with soil particles, and perhaps very slowly biodegraded, if at all.
3. Water-soluble chemicals can get into the air by evapotranspiration (as the water evaporates).
4. Chemicals that are highly volatile can get into the air and not be available for leaching.
5. Chemicals on the soil surface may be phototransformed and thus not be available to leach; however, the photoproducts may be available.
6. The more precipitation, the greater the chance is for chemicals to leach.
7. Hydrolysis and dissociation may prevent leaching.

The reader must remember that any parent compound that breaks down by any mechanism yields breakdown products that may or may not cause problems in the environment. The breakdown products should be regarded in the same light as the parent chemical, even though we are considering *only* parent compounds here for purposes of discussion and simplicity.

There are many ways to study or predict chemical movement through soil, as reported by Helling and Turner [1], Lyman et al. [2], and Ney [3, 4]. They include the use of:

1. Soil columns to measure chemical movement.
2. Solubility data to predict movement (see above).
3. Octanol water partition coefficient to indicate lipo-solubility, which is also an indicator of water solubility, in order to predict movement (see above).
4. Sorption coefficients to predict mobility (see above).
5. Soil TLC to predict mobility (see above) [1].

Table 2-7 shows what can be predicted with a *Koc,* a soil TLC-*Rf,* and water solubility (WS).

Table 2-7. Leaching and Fate of Chemicals.

ENVIRONMENTAL COMPARTMENT: SOIL	KOC > 10,000 Rf < 0.34 WS < 10ppm	KOC 1,000–10,000 Rf 0.34–0.75 WS 10–1,000 ppm	KOC < 1,000 Rf > 0.75 WS > 1,000 ppm
Adsorbs	yes	either way	n*
Leaches	n	either way	yes
Biodegrades	n	either way	yes
Dissipates	n	either way	yes
Accumulates	yes	either way	n
Bioaccumulates	yes	either way	n
Causes food-chain contamination	yes	either way	n

* n denotes negligible.

REFERENCES

[1] Helling, Charles S. and Benjamin C. Turner, "Pesticide Mobility: Determination by Soil Thin-Layer Chromatography," *Science,* vol. 162, pp. 562–563, 1968.

[2] Lyman, Warren J., William F. Reehl, and David H. Rosenblatt, *Handbook of Chemical Property Estimation Methods,* McGraw-Hill Book Company, 1982.

[3] Ney, Ronald E., Jr., "Fate, Transport and Prediction Model Application to Environmental Pollutants," Spring Research Symposium, James Madison University, UP, April 16, 1981.

[4] Ney, Ronald E. Jr., "Exposure Assessment Considerations and Problems," Exposure Assessment Workshop, U.S. Environmental Protection Agency, Washington, D.C., UP, April 6–7, 1982.

RUNOFF

Runoff is the process by which a chemical is moved across a contaminated area to a noncontaminated area. The end result is contamination of the noncontaminated area. This occurs in a number of ways:

1. A water-soluble chemical can be solubilized in rainwater and snow melt, and as the water runs off the contaminated area, so does the soluble chemical.
2. Chemicals that are ad- or absorbed by soil can be moved with the soil as soil erosion occurs.
3. Solvents other than water may solubilize and move chemicals.

Table 2-8. Runoff and Chemical Movement.

SOIL SURFACE	WS < 10 ppm Rf < 0.34 Koc > 10,000 Kow > 1,000	WS 10–1,000 ppm Rf 0.34–0.75 Koc 1,000–10,000 Kow 500–1,000	WS > 1,000 ppm Rf > 0.75 Koc < 1,000 Kow < 500
Water runoff	n*	either way	yes
Soil runoff	yes	yes	yes

* n denotes negligible.

(Remember that solvents other than water are also contaminates, but only water is considered as a solvent herein.)

When a chemical does run off or is moved with soil to a noncontaminated area, contamination of air, soil, plants, animals, and aquatic environmental compartments can result.

There are ways to predict whether runoff could occur. If chemicals are water-soluble, runoff is likely. Land surfaces are not always level; they may have slopes. Therefore, if the land has definite slopes that result in water runoff, it is likely that (1) soil is moved, and (2) any chemicals that are present are moved. One could even "eyeball" the movement of rainwater across land and predict runoff.

Using the WS, *Kow,* soil TLC-*Rf,* or *Koc,* one can predict the movement of a water-soluble chemical across land. Table 2-8 shows how one might predict runoff.

Finally, it should be noted that chemical runoff may occur, to some extent, almost *100%* of the time if water runs off. The extent of the runoff may be small, as the newly contaminated area may be, but movement will occur. Of course, the amount of rainfall and snow strongly infuences the amount and distance of such movement.

Chapter 3
Biological Processes

BIODEGRADATION

Biodegradation is the biological process by which aerobic microbes or anaerobic microbes break down organic chemicals to either a higher- or a lower-molecular-weight chemical(s) called a biodegradate(s). This is a very important process by which soil microbes or aquatic microbes can detoxify chemicals [2]. The process could also result in the formation of a more toxic chemical.

The biodegradation process can be evaluated by hydrolysis, phototransformation, leaching, and sorption studies. There are computer models that predict biodegradation, but at this time they have not been fully validated; and other estimation techniques are in the early stages, as reported by Lyman et al. [1].

Many factors may affect the rate of biodegradation. (Here the rate means the time required to break down the parent chemical through at least one-half of its original concentration.) Among the influences are pH, temperature, sorption, populations of microbes, different types of microbes, moisture, the presence of other chemicals, and the concentration of chemicals present.

Below are some guidelines that can be used to predict whether biodegradation can occur:

1. Chemicals that are highly water-soluble can biodegrade, but those with low WS usually will not.
2. Chemicals that adsorb in soil usually will not biodegrade, but those that do not adsorb can.
3. Chemicals with a high Kow usually will not biodegrade, but those with a low Kow can.

Table 3-1. Biodegradation of Chemicals

SOIL OR WATER	KOC >10,000 RF <0.34 WS<10 PPM KOW >1,000	KOC 1,000–10,000 RF 0.34–0.75 WS 10–1,000 PPM KOW 500–1,000	KOC <1,000 RF >0.75 WS>1,000 PPM KOW <500
Adsorbs	yes	either way	n*
Leaches	n	either way	yes
Soluble	n	either way	yes
Dissipates	n to slowly	either way	yes
Accumulates	yes	either way	n
Bioaccumulates	yes	either way	n
Biodegrades	n to slowly	either way	yes

* n denotes negligible.

4. Chemicals that leach in soil usually will biodegrade, but those that do not leach usually will not.

Table 3-1 shows what can take place.

REFERENCES

[1] Lyman, Warren J., William F. Reehl, and David H. Rosenblatt, *Handbook of Chemical Property Estimation Methods,* McGraw-Hill Book Company, Chapter 9, 1982.

[2] Ney, Ronald E., Jr., "Fate, Transport and Prediction Model Application to Environment Pollutants," Spring Research Symposium, James Madison University, UP, April 16, 1981.

BIOACCUMULATION

If a chemical is taken up by a plant or an animal, the chemical can be metabolized—that is, changed to a higher- or lower-molecular-weight compound that may or may not be more toxic than the parent compound. There are other chemical degradation processes that could change a parent chemical (hydrolysis in animals, etc.), but these processes are not discussed as we only want to predict whether parent chemicals can build up, accumulate, or bioaccumulate in plants or animals.

Bioaccumulation is the key consideration, as we want to be able to predict whether food-chain contamination can occur. It may

Table 3-2. Bioaccumulation of Chemicals

PLANTS AND ANIMALS	WS <10 PPM KOW >1,000 RF<0.34 KOC >10,000	WS 10–1,000 PPM KOW 500–1,000 RF 0.34–0.75 KOC 1,000–10,000	WS>1,000 PPM KOW <500 RF>0.75 KOC <1,000
Bioaccumulation	yes	either way	n*

* n denotes negligible.

happen in two ways: (1) contaminated plants eaten by animals cause contamination in animals, and (2) contaminated animals eaten by animals contaminate animals. This is how contamination is passed up the food chain.

It is important to note that plant and animal metabolism studies are most complex, and studies on the metabolites formed may require new analytical methods and new toxicological studies.

As to the problem of predicting bioaccumulation, some of the predictive indicators are water solubility (WS) and the octanol water partition coefficient (Kow). Table 3-2 indicates how to predict whether bioaccumulation can occur.

Chapter 4
Exposure Assessment

There are many aspects to making exposure assessments. Exposure considerations should be based on the availability of a chemical residue to environmental compartments, with resultant exposure of wildlife and humans (grouped together as animals). Exposure to a chemical depends upon its environmental fate and transport and on how long it lasts in the environment.

The preceding discussions on water solubility, octanol water, hydrolysis, photolysis, volatilization, soil sorption, leaching, run-off, biodegradation, and bioaccumulation indicated whether chemicals would be available for exposure. It should be noted that there are other predictors, but I believe that they are beyond the scope of this discussion and should be left to qualified professionals. Lyman et al. [1], for example, present many other predictive techniques; and Ney [2] discussed considerations and problems associated with exposure assessment of hazardous waste constituents.

Table 4-1 indicates what could happen in the environment. Table 4-2 is a summary of predictive techniques and ranges, and of chemical fate and transport within these ranges, as given in previous chapters.

Exposure to some chemicals may not be harmful. If a chemical is acutely toxic, cancer-causing, mutagenistic, teratogenistic, and so on, then a qualified person should weigh its characteristics against the amount of the chemical available for exposure to animals. Equations and models have been devised for this but they are beyond the scope of this presentation. Remember that the purpose of this book is to help individuals to discern the fate and transport of chemicals in the environment, and to determine whether exposure is possible.

Table 4-1. Processes Affected by Environmental Characteristics.

ENVIRONMENT	PHYSICAL PROCESS	CHEMICAL PROCESS	METABOLIC PROCESS
Wind	Photolysis	Hydrolysis	Plant metabolism
Temperature	Sorption		Animal metabolism
Light	Desorption		Microbial metabolism
Soil	Volatility		
Runoff	Dissociation		
Leaching	Ion-exchange		

Table 4-2. Predicting Where the Chemical Goes.

PREDICTIVE TECHNIQUE	RANGES		
WS, ppm	<10	10–1,000	>1,000
Hydrolysis, $T_{1/2}$ days	>90	30–90	<30
Photolysis, $T_{1/2}$ days	>90	30–90	<30
VP, mm Hg	<0.000001	0.000001–0.01	>0.01
Koc	>10,000	1,000–10,000	<1,000
Rf	<0.34	0.34–0.75	>0.75
Kow	>1,000	500–1,000	<500
Soil, $T_{1/2}$ months	6	2–6	<2
FATE AND TRANSPORT			
Soluble	n*	either way	yes
Hydrolyzes	n	either way	yes
Photolyzes	n*	either way	yes
Volatilizes	n	either way	yes
Adsorbs	yes	either way	n
Leaches	n	either way	yes
Runs off	n	either way	yes
Bioaccumulates	yes	either way	n
Persists	yes	either way	n
Biodegrades	slowly	either way	yes
Is metabolized	slowly	either way	yes

* n denotes negligible.

REFERENCES

[1] Lyman, Warren J., William F. Reehl, and David H. Rosenblatt, *Handbook of Chemical Property Estimation Methods,* McGraw-Hill Book Company, 1982.

[2] Ney, Ronald E., Jr., "Exposure Assessment Considerations and Problems," Exposure Assessment Workshop, U.S. Environmental Protection Agency, Washington, D.C., UP, April 6–7, 1982.

Chapter 5
Examples

This chapter presents some common chemicals along with some characteristic data, followed by a discussion of what may happen to each chemical in the environment, as well as exposure considerations.

There are several things to remember in using this material:

1. Refer to Table 4-2, on making predictions.
2. Remember that there always may be exceptions to the rules suggested by the table.
3. If two or more data points in Table 4-2 are satisfied, then a fit is likely for most points, in any of the three columns.
4. If two or more of a substance's physical, chemical, or biological properties fit a column in the table, then there is good likelihood that most of its other properties will fit the same column.
5. More data are available for most of the chemicals listed than are chosen for the example predictions. Remember that we may need to make predictions based on limited data.

There will be mention in the data, when available, of the bioconcentration factor, BCF, which is a chemical's ability to bioconcentrate or bioaccumulate in animals (fish, in this case). The data will be useful in discussing exposure. Refer to the appendix for an alphabetical listing of all chemicals included in this chapter.

CHEMICALS

Acenaphthene

Data

Pesticide		
WS	3.47 mg/L	[4]
Kow	21,380	[4]
VP at 20°C	0.02 torr	[4]

Discussion

- WS indicates that this chemical could adsorb to soil, run off with soil, and be bioaccumulated, and it should not leach and should not be biodegraded.
- Kow indicates that this chemical should bioaccumulate and could cause food-chain contamination. Residues of this chemical could be expected in the food chain.
- VP indicates that this chemical could be volatile. If it is not phototransformed, fallout of the parent chemical could contaminate the food chain and water, and so could transformation products if they are formed.

Exposure

Routes of exposure could be by (1) ingestion of contaminated food and water and (2) inhalation of the volatile chemical.

Acenaphthylene

Data

WS	3.93 mg/L	[4]
Kow	11,749	[4]
VP	0.029 torr	[4]

Discussion

- WS indicates that this chemical could adsorb to soil, run off with soil, and be bioaccumulated, and it should not leach and should not be biodegraded.

- Kow indicates that bioaccumulation could occur and cause food-chain contamination. Residues of this chemical could be expected in the food chain.
- VP indicates that this chemical could be volatile. If it is not phototransformed, fallout of the parent chemical could contaminate the food chain and water, and so could phototransformation products if they are formed.

Exposure

Routes of exposure could be by (1) ingestion of contaminated food and water and (2) inhalation of volatile residues.

Acephate (O,S-Dimethyl acetylphosphoramidothioate)

Data

Pesticide		
WS	650,000 ppm	[6]
BCF—static water	0	[6]

Discussion

- WS indicates that this chemical should be biodegraded and could leach and run off, and that adsorption and bioaccumulation should not occur.
- BCF indicates that bioaccumulation and food-chain contamination should not occur, and is supported by the WS. BCF indicates that this chemical may be metabolized in animals.

Exposure

The routes of exposure could be by (1) inhalation if the chemical were volatile (but no data are presented on volatility), and (2) by ingestion of contaminated water and food if that occurred. Additional data would be needed for a more conclusive prediction.

Acetophenone

Data

Use—perfumery		[8]
Water solubility (WS)	5,500 ppm	[3]
Octanol water partition (Kow)	38.6 ± 1.2	[3]
Soil sorption coefficient (Koc)	35	[3]

Discussion

- WS is high; thus mobility should occur with water, and the chemical should be biodegraded by microbes.
- Kow is low, which indicates that bioaccumulation should not occur, which also is indicated by the high WS.
- Koc is low, which indicates that soil adsorption should not occur and that accumulation in soil should not occur.
- The three characteristics indicate that acetophenone should not be persistent in the environment. Bioaccumulation and food-chain contamination should not occur, based on the WS and the Koc. The WS, Koc, and Kow indicate that the chemical could leach and run off into water supplies if in soils that have no or limited microbial populations for biodegradation.

Exposure

Immediate exposure could be a problem if this chemical were acutely toxic. This chemical, if not degraded, could leach or run off into the aquatic environment and contaminate drinking water. Therefore, ingestion of contaminated drinking water could be a problem. If it were volatile, inhalation could be a problem, but no data are presented herein on volatility.

Acridine

Data

Use—manufacture of dyes		[3]
WS	38 ppm	[3]
Kow	4,200 ± 940	[3]
Koc	12,910	[3]
Volatile with steam		[3]

Discussion

- WS indicates that acridine should not be mobile (leached) and should not be biodegraded by microbes.
- Kow indicates that bioaccumulation and food-chain contamination are likely.
- Koc indicates that soil adsorption should occur, and this is supported by the low WS.
- Based on these aspects, the chemical acridine should not leach, could accumulate in soil, and could bioaccumulate and cause food-chain contamination. Runoff could occur with soil particles, resulting in aquatic contamination.

Exposure

Exposure could be through food-chain consumption. If the chemical causes cancer through long-term exposure, this could be a problem as long as the food is consumed. If it is in drinking water, the same exposure problem could exist. The chemical is volatile in steam and not at ambient temperatures; thus inhalation of this chemical should not be a problem at ambient temperatures.

Acrylonitrile (also Vinyl cyanide)

Data

WS	73,500 mg/L	[4]
VP at 22.8°C	100 torr	[4]
Kow	1.38	[4]
Degraded by microbes		[4]

Discussion

- WS indicates that runoff, leaching, and biodegradation (as reported) could occur. The chemical should not bioaccumulate or adsorb in soil.
- VP indicates volatility; thus contamination to water, crops, and animals could occur via fallout. Inhalation could be a problem.
- Kow indicates that bioaccumulation should not occur.
- Biodegradation by microbes could occur.

Exposure

Routes of exposure could be by (1) inhalation of volatile residue; (2) ingestion of contaminated water, plants, and animals if the chemical precipitated out of the atmosphere; and (3) ingestion of water if the chemical leached to the drinking water supply.

Alachlor (2-Chloro-2′,6′-diethyl-*N*-(methoxymethyl) acetanilide) (also Lasso)

Data

Pesticide		
WS	242 ppm	[6]
Koc	190	[6]
Kow	830	[6]
BCF—static water	0	[6]

Discussion

- WS indicates that this chemical could be either mobile or not, be bioaccumulated or not, and be adsorbed or not. It is predicted that the chemical should be mobile.
- Koc indicates some adsorption.
- Kow indicates that bioaccumulation is possible.
- BCF indicates that bioaccumulation should not occur, and that metabolism in animals could occur, based on the zero BCF number.
- The chemical could be mobile and get into water. Biodegradation may occur in soils and prevent leaching; however, no data are presented to support this prediction.

Exposure

Exposure could be by routes of (1) drinking contaminated water if mobility occurred, (2) inhalation if the chemical were volatile, and (3) ingestion of contaminated food if that occurred. No data are presented to support these predictions.

Aldicarb (2-methyl-2-(methylthio) propionaldehyde O-(methylcarbanoyl) oxime)

Data

Pesticide		
WS	7,800 ppm	[6] [9]
Kow	11.02	[9]
Koc	0.073	[9]
BCF—static water	42	[6]

Discussion

- WS indicates that this chemical could leach into groundwater (it has done so) and could run off. It could be biodegraded in soils with good microbial populations and high organic matter contents.
- Kow indicates that this chemical should not bioaccumulate, and this is supported by the high WS.
- Koc indicates this chemical should not adsorb to soil, and this is supported by the high WS and low Kow.
- BCF is indicative that residues may be found in fish, but bioaccumulation should not occur based on the BCF number and the Kow number. This chemical is acutely toxic; thus fish would probably die. Ingestion of contaminated drinking water could be a problem.

Exposure

Immediate exposure is a problem, as this chemical is acutely toxic. If drinking water is contaminated, one must consider what is a safe level for drinking and what levels may be a danger to aquatic organism and animals that eat the organisms (dead or alive). If this chemical reaches soils with low organic matter content and low microbial populations, leaching could occur and has occurred. Inhalation could be a problem if this chemical were found to be volatile, but no volatility data are presented herein.

Aldrin (Hexachlorohexahydro-endo, exo-dimethanonaphthalene 95% and related compounds 5%)

Data

Pesticide		
WS	0.013 ppm	[6]
Koc	410	[6]
BCF—static water	3,140	[6]

Discussion

- WS indicates that leaching and biodegradation should not occur, and that adsorption and bioaccumulation could occur. Runoff with soil particles could occur.
- Koc indicates that adsorption should not occur to any great extent, but it does occur, although perhaps not enough to prevent leaching to groundwater. The Koc value is contradictory to WS and BCF values, which indicate adsorption to soil.
- BCF indicates bioaccumulation and food-chain contamination. Residues could be expected in the food chain.
- WS and BCF indicate soil adsorption and bioaccumulation, as well as the potential for residues to be in the food chain. Koc indicates just the opposite, and indicates potential for leaching. Additional data would be needed to give a more precise prediction. However, two values do support each other and can be used until proven otherwise (i.e., the WS and BCF values).

Exposure

Routes of exposure could be by (1) ingestion of contaminated food and water and (2) ingestion of contaminated food. If the chemical were volatile, inhalation could be a problem; however, no data are presented herein on volatility. Food-chain contamination could be a major problem.

Amitrole (3-Amino-s-triazole) (also 3-AT)

Data

Pesticide		
Soil TLC-Rf	0.73	[1]

Discussion

- Soil TLC indicates leaching; thus this chemical should be water-soluble, biodegraded, and mobile, and it should not be adsorbed or bioaccumulated.
- With only one piece of data any prediction could be entirely wrong; however, when a prediction is required, one must be given with explanations and disclaimers as needed. In this case, additional data would be needed to support any predictions.

Exposure

Routes of exposure could be by (1) drinking contaminated water and (2) ingestion of contaminated food. If the chemical were volatile, inhalation could be a problem; however, no data are presented herein on volatility. Additional data would be needed to support and verify any of these predictions.

Ammonia

Data

Used in refrigeration and many other uses	[8]
Gas	[8]

Discussion

- If ammonia gas were released into the environment, inhalation could be a problem. Such releases could occur from refrigeration units that use ammonia, manufactures, industries, and water purification processing. Phototransformation may be a problem; however, there is no indication of such.

Exposure

Route of exposure could be by inhalation of volatile residues.

Aniline

Data

Pesticide	
Used in manufacturing of dyes, perfumes, and medicinals	[8]

WS	36,600 ppm	[6]
Kow	7	[6]
BCF—static water	6	[6]

Discussion

- WS indicates that this chemical could leach, run off, and be biodegraded, and it should not adsorb to soil and should not be bioaccumulated.
- Kow indicates that this chemical should not bioaccumulate.
- BCF indicates that bioaccumulation should not be a problem; however, there is a possibility of residues in the food chain. The possibility of metabolism in animals also is indicated, based on the low BCF value.

Exposure

Routes of exposure could be by (1) ingestion of contaminated food and water if such occurred, and (2) inhalation if the chemical were volatile, but no data on volatility are presented herein.

Anthracene (also Paranaphthalene)

Data

Pesticide

WS	0.073 ppm	[6]
Koc	26,000	[6]
Kow	22,000	[6]

Discussion

- WS indicates that leaching and biodegradation should not occur, and that adsorption, bioaccumulation, and runoff with soil particles could occur.
- Koc indicates adsorption to soil, and this should prevent leaching.
- Kow indicates that bioaccumulation and food-chain contamination could occur. This prediction is supported by WS and Koc values.

Exposure

Exposure routes could be by (1) ingestion of contaminated food and water and (2) inhalation if volatility occurred; however, no data are presented on volatility. If this chemical got into the aquatic environment, it could bioaccumulate and cause food-chain contamination. There are indications that consumption of contaminated water and food could result in chemical residues in animals. If it is also possible that bioaccumulation in plants could occur, and consumption of contaminated plants could also cause food-chain contamination.

Asbestos

Data

Naturally occurring and has many uses	
WS—not soluble	[4]
Suspended in water	[4]
Not biodegraded	[4]
Not chemically degraded	[4]
Not volatile	[4]
Picked up by air as an aerosol	[4]
Not photolyzed	[4]
Can absorb other chemical	[4]

Discussion

- Asbestos is stable in the environment; therefore, it is resistant to chemical breakdown, biodegradation, phototransformation, and probably metabolism. Dissipation is unlikely in the environment.
- Asbestos fibers can be physically broken down to smaller fibers.
- Asbestos can be airborne; therefore, inhalation of the fibers is a problem. Airborne asbestos fibers can fall out to contaminate water, and thus be suspended in the water; they should not volatilize or break down in water, but should be persistent in water, and could be a problem in fish gills. Fallout of airborne asbestos fibers can also contaminate surfaces, such as food.

Exposure

Routes of exposure could be by (1) consumption of contaminated water with suspended asbestos particles, (2) consumption of food with asbestos on its surface, and (3) inhalation of asbestos air-borne fibers.

Atrazine (2-Chloro-4-ethylamino-6-isopropylamino-s-triazine)

Data

Pesticide

WS	33 ppm	[2] [6]
VP	1.4 mm Hg $\times$ 10^{6} at 30°C	[2]
Koc	149	[6]
Kow	476	[6]
BCF—flowing water	0	[6]
BCF—static water	11	[6]
Biodegraded by microbes		[2]
Soil TLC-Rf	0.47; class 3	[5]

Discussion

- WS indicates some solubility; thus biodegradation may occur, and this has been reported.
- VP indicates volatility.
- Koc indicates that adsorption should not occur in soil.
- Kow indicates that bioaccumulation should not occur.
- Soil TLC-Rf of 0.47 in class 3 does indicate that mobility is possible; thus the WS of 33 ppm is supported by mobility class 3. This chemical may get into water.
- Based on these data, volatility may be of more concern than mobility through or across soils.

Exposure

Exposure could occur if this chemical reached groundwater, or surface water, resulting in the contamination of drinking water.

Biodegradation may be too slow to prevent leaching, and adsorption to soil should not occur. If this chemical finds its way to sandy soils, low-organic soils, and soils with low microbial content, then leaching is likely. In any case, the toxicity of this chemical may be of concern, and that should be looked into if drinking water is contaminated. Because of its volatility, inhalation exposure should be of concern. Fallout of volatile residues into noncontaminated areas could be a problem unless Atrazine is phototransformed; thus food-chain contamination is possible.

Benefin (also *N*-Butyl-*N*-ethyl-alpha, alpha, alpha-trifluoro-2, 6-dinitro-*p*-toluidine)

Data

Pesticide		
WS	<1 ppm	[6]
Koc	10,700	[6]

Discussion

- WS indicates that benefin should not leach in soil, but could run off with soil particles.
- Koc indicates that benefin should accumulate, and adsorb in soils.
- Based on the WS and the Koc, benefin may bioaccumulate in the food chain and not be or only slowly be biodegraded.
- If this chemical gets into water, it may be persistent, and it could bioaccumulate, resulting in food-chain contamination.

Exposure

Exposure could be by crop uptake, if crops are grown in contaminated soils, with the chemical passed on to animals that eat the food. Were the chemical to get into water, food-chain contamination could occur. From these two pieces of data, exposure would be by ingestion, and this would not preclude other exposure routes if more data were given, such as inhalation.

Bentazon (3-Isoprophyl-1*H*-2,1,3-benzothiadizen-4 (3*H*)-one 2,2-dioxide) (also Basagran)

Data

Pesticide		
WS	500 ppm	[6]
Koc	0	[6]
Kow	220	[6]
BCF—static water	0	[6]

Discussion

- WS indicates that the chemical could run off, leach, and be biodegraded, and it should not bioaccumulate or adsorb to soil.
- Koc indicates that adsorption should not occur.
- Kow indicates that the chemical should not bioaccumulate, and food-chain contamination should not occur.
- BCF indicates that bioaccumulation should not occur, and that residues should not be expected in animals. The BCF value of zero indicates that the chemical is metabolized in animals. Koc and Kow values support these predictions.

Exposure

Routes of exposure could be by (1) ingestion of contaminated water, if such occurred; (2) inhalation, if the chemical were volatile, but no data on volatility are presented herein; and (3) ingestion of contaminated food, if such occurred.

Benzene

Data

WS	1,780 ppm	[6]
Koc	83	[6]
Kow	135	[6]
VP	95.2 torr	[4]
VP	76 mm Hg	[7]

Discussion

- WS indicates that benzene could be mobile in soils, and may be biodegraded.
- Koc indicates that adsorption should not occur.
- Kow indicates that bioaccumulation should not occur.
- VP indicates that benzene could volatilize and get into air. Volatilization should help to prevent leaching in soils.
- Benzene could leach, volatilize, and be biodegraded by microbes. Bioaccumulation should not be a problem; thus food-chain contamination is unlikely. Inhalation could be a problem.

Exposure

Exposure routes can be predicted to be (1) inhalation due to volatilization and (2) drinking contaminated water if benzene gets to drinking water. Fallout of volatile residues could be a problem if benzene is not phototransformed.

Benzidine

Data

WS	400 mg/L	[4]
Adsorbs to clay particles		[4]
Absorbs light	287 to 340 nm	[4]
Kow	64	[4]

Discussion

- WS indicates that the chemical could run off, leach, and be biodegraded.
- This chemical adsorbs to clay particles and not to organic matter (which is an unusual process, but still adsorption). This could not be predicted based on the data used herein.
- The chemical absorbs light, which indicates that the chemical could be phototransformed to other products.
- Kow indicates that bioaccumulation is unlikely.

Exposure

Routes of exposure could be by drinking contaminated water if leaching and runoff occurred. It cannot be predicted whether adsorption to clay is significant enough to prevent leaching because no data are presented. Inhalation could be a problem, if the chemical were volatile, as could phototransformation products; however, no data are presented to make such predictions.

Benzo [*a*] anthracene

Data

WS at 25°C	0.014 mg/L	[4]
Kow	407,380	[4]
VP at 20°C	5×10^{-9} torr	[4]

Discussion

- WS indicates that this chemical could adsorb to soil, run off with soil, and be bioaccumulated, and it should not leach and should not be biodegraded.
- Kow indicates bioaccumulation and potential for food-chain contamination. Residues could be expected in the food chain.
- VP indicates that this chemical should not be volatile.

Exposure

Routes of exposure could be by ingestion of contaminated food and water if this chemical were released in the environment.

Benzo [*b*] fluoranthene

Data

WS at 25°C	0.012 mg/L	[4]
Kow	3,715,352	[4]
VP at 20°C	5×10^{-7} torr	[4]

Discussion

- WS indicates that this chemical could absorb to soil, run off with soil, and be bioaccumulated, and it should not leach and should not be biodegraded.
- Kow indicates bioaccumulation and potential for food-chain contamination. Residues could be expected in food and water.
- VP indicates that this chemical should not be volatile.

Exposure

Routes of exposure could be by ingestion of contaminated food and water if this chemical were released into the environment.

Benzo [*a*] pyrene

Data

WS at 25°	0.0038 mg/L	[4]
Kow	1,096,478	[4]
VP at 20°	5×10^{-9} torr	[4]

Discussion

- WS indicates that the chemical could adsorb in soil, run off with soil, and be bioaccumulated, and it should not leach and should not be biodegraded.
- Kow indicates bioaccumulation and potential food-chain contamination. Residues could be expected in food and water.
- VP indicates that volatility is of no concern.

Exposure

Routes of exposure could be by ingestion of contaminated food and water.

Bifenox (Methyl 5-(2,4-dichlorophenoxy)-2-nitrobenzoate) (also Modown)

Data

WS	0.35 ppm	[6]
BCF—static water	200	[6]

Discussion

- WS indicates that this chemical should adsorb to soil, run off with soil, and be bioaccumulated, and it should not leach and should not be biodegraded.
- BCF indicates that this chemical could bioaccumulate and could cause food-chain contamination. Residues could be expected in the food chain if the chemical were released in the environment.

Exposure

Routes of exposure could be by (1) consumption of contaminated water and food if such occurred, and (2) inhalation if the chemical were volatile; however, no data are presented herein on volatility.

Biphenyl (also Diphenyl)

Data

Pesticide

WS	7.5 ppm	[6]
Kow	7,540	[6]
BCF—flowing water	340	[6]

Discussion

- WS indicates that this chemical should not leach and not be biodegraded; however, it could adsorb to soil, run off with soil, and be bioaccumulated.
- Kow indicates that bioaccumulation should occur.
- BCF indicates that a residue could be present, and that this chemical should not bioaccumulate. This contradicts what the WS and Kow values indicate. BCF indicates that residues could be found in animals; however, metabolism should reduce the amount of residue found.
- WS and Kow are in contradiction with BCF, and this indicates that additional testing is needed; however, BCF does indicate that there may be a problem with residues in the food chain, though not bioaccumulated. Metabolism of the chemical in animals could occur, but it might be at a very slow rate.

Exposure

Routes of exposure could be by (1) ingestion of crops, animals, and water if they were contaminated, and (2) inhalation of volatile residue if such occurred; however, no data are presented on volatility.

Bromacil (5-Bromo-3-*sec*-butyl-6-methyluracil)

Data

Pesticide		
WS	8.5 ppm	[6]
Koc	72	[6]
Soil TLC-Rf	0.69	[1]

Discussion

- WS indicates that the chemical should be adsorbed to soil, run off with soil particles, and be bioaccumulated, and it should not be biodegraded and should not be leached.
- Koc indicates that adsorption to soil should not occur; however, it could occur in small amounts.
- Soil TLC-*Rf* indicates that the chemical could leach.
- The three pieces of data are contradictory; however, Koc and soil TLC indicate that leaching could occur, and that is what should be predicted.

Exposure

Routes of exposure could be by (1) drinking contaminated water, (2) ingestion of contaminated food, and (3) inhalation if volatility occurred; however, no data are presented herein to indicate volatility. Because of the contradiction betweem WS, Koc, and soil TLC, additional data would be needed for a more conclusive prediction of what could happen to this chemical in the environment.

Bromobenzene

Data

WS	446 ppm	[6]
Kow	900	[6]

Discussion

- WS indicates that this chemical could go either way in relation to leaching, runoff, adsorption, degradation, and bioaccumulation.
- Kow indicates that this chemical could bioaccumulate or be slowly metabolized. Residues could be expected in the food chain.
- There are not enough data for one to make a precise prediction with the data given. However, a good prediction, using the limited data given, would be that the chemical could result in the contamination of water and the food chain, and residues could be expected in the food chain.

Exposure

Routes of exposure could be by (1) consumption of contaminated water and food if such occurred, and (2) inhalation if the chemical were volatile; however, no data are presented herein on volatility.

Bromodichloromethane

Data

Kow	76	[4]
VP at 20°C	50 torr	[4]

Discussion

- Kow indicates that this chemical should not bioaccumulate; thus it is also predicted that it could leach, run off, and be biodegraded, and it should not be adsorbed.
- VP indicates that the chemical should be volatile; thus inhalation could be a problem. If it were phototransformed, photoproducts and the parent chemical could fall out and contaminate water and food.
- This chemical could be mobile in the environment, and could contaminate water and food, and thus could result in contamination of the food chain. Volatile residues could present an inhalation problem.

Exposure

Routes of exposure could be by (1) consumption of contaminated water and food if such occurred, and (2) inhalation of volatile residues.

Bromoform (also Tribromomethane)

Data

WS at 30°C	3,190 mg/L	[4]
Kow	200	[4]
VP at 34°C	10 torr	[4]

Discussion

- WS indicates that this chemical should leach, run off, and be biodegraded, and it should not adsorb and should not be bioaccumulated.
- Kow indicates that the chemical should not bioaccumulate, and this is supported by the value given for WS.
- VP indicates that volatility should occur for this chemical. If it were phototransformed, both the parent chemical and photoproducts could fall out and contaminate water and food. Inhalation could be a problem.
- This chemical is mobile and could contaminate water and food, thus resulting in contamination of the food chain. Residues in the environment could dissipate by biodegradation and metabolism; however, no data are presented on these mechanisms. Inhalation of volatile residues could be a problem.

Exposure

Routes and exposure (1) could be by consumption of contaminated water and food, if such occurred, and (2) should be by inhalation of volatile residues.

Bromomethane (also Methyl bromide)

Data

WS	900 mg/L	[4]
Kow	12.59	[4]
VP at 20°C	1,420 torr	[4]

Discussion

- WS indicates that this chemical could go either way, but Kow indicates that it should not bioaccumulate; therefore, it is predicted that it should leach, run off and biodegrade, and it should not adsorb to soil and should not be bioaccumulated.
- Kow indicates that bioaccumulation should not occur. It is possible that residues of this chemical could be found in water and food; however, its volatility should help prevent this.
- VP indicates that this chemical is volatile, and this could cause inhalation problems. If it were phototransformed, fallout of photoproducts and the parent chemical could contaminate the environment.
- This chemical is volatile and could present inhalation problems with the parent chemical and potential photoproducts. Although residues could be present in the environment, it is predicted that they would not be there long because they would volatilize.

Exposure

Route of exposure should be by inhalation of volatile residues.

4-Bromophenyl phenyl ether

Data

WS at 20°C	38 mg/L	[4]
Kow	141,254	[4]
VP at 20°C	0.0015 torr	[4]

Discussion

- WS indicates that this chemical could adsorb to soil, run off with soil, and be bioaccumulated, and that it should not leach and should not be biodegraded.
- Kow indicates bioaccumulation and potential for food-chain contamination. Residues could be expected in the food chain.
- VP indicates potential for volatility. If volatility occurred, fallout could cause contamination of water and the food chain. If the chemical were phototransformed, transformation products also could contaminate the food chain and water.

Exposure

Routes of exposure could be by (1) ingestion of contaminated food and water and (2) inhalation of volatile residues.

Bufencarb (3- (1-Methylbutyl) phenyl methylcarbamate) (also Bux, Metalkamate, etc.

Data

Pesticide		
WS	<50 ppm	[6]
BCF—static water	0	[6]

Discussion

- WS indicates that this chemical could go either way in relation to leaching, runoff, adsorption, biodegradation, and bioaccumulation.
- BCF indicates that the chemical should not bioaccumulate.
- This chemical should not bioaccumulate, but could be mobile enough to contaminate the food chain. If this occurred, it should be metabolized and biodegraded. There is a possibility that residues could exist in the food chain if dissipation were not rapid enough.

Exposure

Routes of exposure could be by (1) consumption of contaminated water and food if such occurred, and (2) inhalation if the chemical were volatile; however, no data are presented herein on volatility.

Butralin (4-(1,1-Dimethyl) -N-(1-methylpropyl) -2, 6-dinitrobenzenamine) (also Sector)

Data

WS	1 ppm	[6]
Koc	8,200	[6]

Discussion

- WS indicates that this chemical should be adsorbed to soil, and be bioaccumulated and runoff with soil, and it should not leach and should not be biodegraded.
- Koc indicates that this chemical should be adsorbed to soil and could be bioaccumulated; this is supported by the water solubility value.

Exposure

Routes of exposure could be by (1) consumption of contaminated water and food if such occurred, and (2) inhalation if the chemical were volatile; however, no data are presented herein on volatility

Butyl benzyl phthalate

Data

WS	insoluble	[4]
Kow	26,301	[4]

Discussion

- WS indicates that this chemical could bioaccumulate, adsorb to soil and run off with soil particles, and it should not be biodegraded and should not be leached.
- Kow indicates that it could bioaccumulate and cause food-chain contamination.

Exposure

Routes of exposure could be by (1) ingestion of contaminated food or water and (2) inhalation if the chemical were volatile; however, there are no data presented to indicate volatility.

x-*sec*-Butyl-4-chlorodiphenyloxide

Data

WS	0.14 ppm	[6]
Kow	16,000	[6]
BCF—flowing water	298	[6]
BCF—static water	400	[6]

Discussion

- WS indicates that this chemical should adsorb in soil, run off with soil, and be bioaccumulated, and it should not be leached and should not be biodegraded.
- Kow indicates that the chemical should be bioaccumulated and could cause contamination to the food chain.
- Both BCF values indicate that the chemical should be in the food chain; however, bioaccumulation may be prevented by metabolism in animals. No matter which occurs, high residues should be expected in animals.
- This chemical should be expected to be present in the food chain if released into the environment. Residues also should be expected in the food chain; however, they may be slowly metabolized in animals.

Exposure

Routes of exposure (1) should be by consumption of contaminated water and food if such occurred, and (2) could be by inhalation if the chemical were volatile; however, no data are presented herein on volatility.

Captan (N-trichloromethylthio-4-cyclohexene-1,2-dicarboximide) (also Orthocide and Vancide 89)

Data

Pesticide		
WS	<0.5 ppm	[6]
Kow	224	[6]
BCF—static water	0	[6]
Soil TLC-Rf	0.39	[1]

Discussion

- WS indicates that this chemical should be adsorbed in soil, be bioaccumulated, and run off with soil, and it should not leach and should not be biodegraded.
- Kow indicates that bioaccumulation should not occur.
- BCF indicates that bioaccumulation should not occur, and that metabolism in animals should occur.

- Soil TLC-Rf indicates that there is a possibility of leaching. There is a contradiction between WS, and Kow, BCF, and soil TLC. However, one could predict that leaching is possible, and that residue could be found in the food chain.

Exposure

Routes of exposure could be by (1) drinking contaminated water, (2) ingestion of contaminated food, and (3) inhalation if the chemical were volatile; however, no data are presented on volatility herein.

Carbaryl (1-Naphthyl methylcarbamate) (also Sevin)

Data

Pesticide		
WS	40 ppm	[6]
Koc	230	[6]
Kow	230	[6]
Soil TLC-Rf	0.38	[1]
BCF	<1	[6]

Discussion

- WS indicates that leaching and runoff could occur.
- Koc shows that adsorption should not occur.
- Kow shows that accumulation should not occur.
- Soil TLC-Rf indicates that leaching could occur but should be slight.
- BCF indicates that food-chain bioaccumulation should not occur.
- Carbaryl should biodegrade based on solubility, could leach and run off, but should not accumulate in soil or bioaccumulate; thus food-chain contamination is unlikely.

Exposure

Exposure would be by direct contact or by consumption of contaminated food or contaminated drinking water if carbaryl reached drinking water prior to biodegradation.

Carbofuran (2,3-Dihydro-2,2-dimethyl-7-benzofuranyl methylcarbamate) (also Furadan)

Data

Pesticide

WS	415 ppm	[6]
Kow	40	[6]
BCF—static water	0	[6]

Discussion

- WS indicates that the chemical could leach, run off, and be biodegraded, and it should not be adsorbed and should not be bioaccumulated.
- Kow indicates that bioaccumulation is unlikely.
- BCF indicates that bioaccumulation is unlikely.
- This chemical has caused bird kills; thus it may cause death in some animals prior to any bioaccumulation.

Exposure

Routes of exposure could be by (1) drinking contaminated water, (2) ingestion of contaminated food, and (3) inhalation if the chemical were volatile; however, no data on volatility are presented herein. This chemical has been known to cause bird kills; thus residues in the food chain are possible.

Carbon tetrachloride (or Tetrachloroethane)

Data

WS	785 ppm	[4]
VP at 20°C	90 torr	[4]
Hydrolysis	$t_{1/2}$ 7,000 years	[4]
Absorbs short wavelengths of light and higher-energy UV in the stratosphere		[4]

Discussion

- WS indicates that leaching, runoff, and biodegradation could occur.

- VP indicates that volatility could occur.
- Hydrolysis indicates that this chemical should not hydrolyze in water to be of any degradation benefit. It should be stable in environments without microbes.
- Phototransformation is unlikely to occur in the breathing atmosphere; however, it could occur in the stratosphere.
- Carbon tetrachloride could be available for inhalation and be mobile enough to get into water if biodegradation is slow. Bioaccumulation is unlikely, as is food-chain contamination.

Exposure

Exposure could be by two routes: (1) inhalation due to the chemical's volatility and (2) drinking contaminated drinking water. Acute dangers due to inhalation should be evaluated. Fallout of volatile residues to noncontaminated areas could be a problem, especially to aquatic environments, as this chemical should not be phototransformed.

Chloramben (3-amino-2,5-dichlorobenzoic acid) (also Amiben)

Data

Pesticide		
WS	700 ppm	[6]
Koc	21	[6]

Discussion

- WS indicates that this chemical could go either way in relation to leaching, adsorption, runoff, biodegradation, and bioaccumulation.
- Koc indicates that adsorption is negligible; therefore, it is predicted that this chemical could leach, runoff and be biodegraded, and should not be absorbed and not be bioaccumulated.

Exposure

Routes of exposure could be by (1) consumption of contaminated water and food if such occurred, and (2) inhalation if the chemical were volatile; however, no data are presented herein on volatility. Additional data are needed for more precise prediction.

Chlorbromuron (3-(4-bromo-3-chlorophenyl)-1-methoxy-1-methylurea)

Data

Pesticide		
WS	50 ppm	[6]
Koc	460	[6]
Soil TLC-Rf	0.14	[1]

Discussion

- WS indicates that this chemical could leach, run off, and be biodegraded, and it should not adsorb to soil and should not be bioaccumulated.
- Koc indicates that adsorption to soil should not occur; however, some adsorption is possible.
- Soil TLC-Rf indicates that leaching is unlikely, which contradicts the WS; however, perhaps biodegradation had occurred in this case, or maybe adsorption had taken place. Remember that the WS is in the "either way" range.
- Bioaccumulation of residues in the food chain may be possible, based on Koc and soil TLC values.

Exposure

Routes of exposure could be by (1) drinking contaminated water, (2) ingestion of contaminated food, and (3) inhalation if the chemical were volatile; however, no data are presented herein to indicate volatility.

Chlordane, Technical (60% Octachloro-4,7-methanotetrahydroindane and 40% related compounds)

Data

Pesticide—termite control		
WS	0.056 ppm	[6]
BCF—flowing water	11,400	[6]
BCF—static water	3,140	[6]
Hydrolysis—loses Cl in presence of alkaline reagents		[8]

Discussion

- WS indicates that chlordane should not be biodegraded in soils, should adsorb to soil, should accumulate in soil, and should be persistent in the soil environment.
- BCF indicates that it should bioaccumulate and cause food-chain contamination.
- Hydrolysis is likely in alkaline soils.
- Chlordane is persistent and can bioaccumulate in the environment and be taken up by plants and animals. It could persist in water, and if it gets into the aquatic environment, food-chain contamination is likely.

Exposure

Exposure could be by (1) drinking contaminated water or (2) eating contaminated plants and animals. If this chemical were volatile, than fallout to noncontaminated areas would be of concern, as well as phototransformation. Inhalation of volatile residues could then be a problem. Bioaccumulation and food-chain contamination are likely if this chemical is released to the environment.

Chlorine

Data

Used in water purification and many other uses	[8]
Gas	[8]

Discussion

- If released, chlorine could cause an inhalation problem. It also can combine with other chemicals in the air to form acids, which may present environmental problems.

Exposure

Routes of exposure could be by (1) inhalation of volatile residues and (2) inhalation of acids that might be formed in air.

Chlorobenzene

Data

WS at 20°C	almost insoluble	[4]
Kow	692	[4]
VP at 20°C	8.8 torr	[4]

Discussion

- WS indicates that this chemical could adsorb to soil, run off with soil, and be bioaccumulated, and it should not leach and should not be biodegraded.
- Kow indicates that perhaps bioaccumulation could occur, and residues could be expected in the food chain.
- VP indicates volatility, and fallout of volatile chemicals could contaminate food and water. If the chemical were phototransformed, transformation products could contaminate food and water. Inhalation of the parent compound and its potential photoproducts could be a problem.

Exposure

Routes of exposure could be by (1) ingestion of contaminated food and water if such occurred, and (2) inhalation of volatile residues.

4-Chlorobiphenyl

Data

WS	1.65 ppm	[6]
Kow	79,400	[6]
BCF—flowing water	590	[6]

Discussion

- WS indicates that this chemical could be adsorbed to soil, run off with soil, and be bioaccumulated, and it should not leach and should not be biodegraded.
- Kow indicates that this chemical should bioaccumulate and cause food-chain contamination.

- BCF indicates that this chemical should bioaccumulate. Residues could be expected in the food chain if contamination occurred. The chemical appears to have potential for metabolism in animals; however, this should be very slow, based on the BCF value.
- If this chemical were to get into water, food-chain contamination could be expected.

Exposure

Routes of exposure could be by (1) consumption of contaminated water and food if such occurred, and (2) volatility if the chemical were volatile; however, no data are presented herein on volatility.

p-Chloro-*m*-cresol

Data

Antiseptic and disinfectant		[8]
WS at 20°C	3,850 mg/L	[4]
Kow	891	[4]
Aqueous solution turns yellow when exposed to light and air		[8]
Volatile in steam		[8]

Discussion

- WS indicates that this chemical could leach, run off, and be biodegraded, and it should not adsorb in soil and should not be bioaccumulated.
- Kow indicates that maybe bioaccumulation could occur, which contradicts WS. In either event, residues could be expected in the food chain.
- Exposure of this chemical in aqueous solution to light and air could transform it to other products.
- Because the chemical is volatile in steam, volatility would not be expected at ambient temperatures.

Exposure

Routes of exposure could be by (1) consumption of contaminated drinking water if this chemical leached or ran off into water, and (2) ingestion of contaminated food.

4-Chlorodiphenyloxide

Data

WS	3 ppm	[6]
Kow	12,000	[6]
BCF—flowing water	736	[6]

Discussion

- WS indicates that this chemical could be adsorbed to soil, run off with soil, and be bioaccumulated, and it should not be leached and should not be biodegraded.
- Kow indicates that this chemical should bioaccumulate and cause contamination in the food chain.
- BCF indicates that this chemical should bioaccumulate, and that residues could be expected in the food chain. It may be metabolized in animals; however, metabolism should be very slow.
- This chemical should bioaccumulate and cause food-chain contamination. Residues could be expected in the food chain.

Exposure

Routes of exposure could be by (1) consumption of contaminated water and food if such occurred, and (2) inhalation if the chemical were volatile; however, no data are presented herein on volatility.

Chloroethane (also Ethyl chloride)

Data

WS	5,740 mg/L	[4]
Kow	34.67	[4]
VP at 20°C	1,000 torr	[4]

Discussion

- WS indicates that this chemical should leach, run off, and be biodegraded, and it should not adsorb to soil and should not be bioaccumulated.
- Kow indicates that the chemical should not bioaccumulate, and this is in agreement with the value for WS.
- VP indicates volatility, and this could cause inhalation problems. If the chemical were phototransformed, photoproducts and the parent chemical could fall out and contaminate water and food.
- The chemical should leach and run off into aquatic environments; however, this may not be a problem because of its high volatility and potential to biodegrade. Inhalation of the volatile parent chemical and potential photoproducts should be the major concern. Contamination of the food chain is unlikely.

Exposure

Route of exposure should be inhalation of volatile residues.

Bis (2-Chloroethoxy) methane

Data

WS	81,000 mg/L	[4]
Kow	18	[4]
VP at 20°C	<0.1 torr	[4]

Discussion

- WS indicates that this chemical could leach, run off, and be biodegraded, and it should not adsorb in soil and should not be bioaccumulated.
- Kow indicates that bioaccumulation should not occur; however, if residues did get into the environment, some residues could be expected in food and water.
- VP indicates volatility, and fallout could contaminate food and water. If the chemical were phototransformed, transformation products also could contaminate food and water via fallout.

Inhalation of the parent chemical and its potential photoproducts could be a problem.

Exposure

Routes of exposure could be by (1) ingestion of contaminated food and water and (2) inhalation of a volatile residue.

Bis (2-Chloroethyl) ether

Data

WS	10,200 mg/L	[4]
Kow	38	[4]
VP at 20°C	0.71 torr	[4]

Discussion

- WS indicates that this chemical could leach, run off, and be biodegraded, and it should not adsorb to soil and should not be bioaccumulated.
- Kow indicates that the chemical should not bioaccumulate, and this is supported by the value for WS.
- VP indicates volatility, and thus the potential for inhalation problems. If this chemical is phototransformed, photoproducts and the parent chemical should fall out onto water and food, causing food-chain contamination.
- This chemical could contaminate the environment by leaching, runoff, and volatility. No data have been presented to indicate if biodegradation is rapid enough to prevent residues from occurring in the environment. Inhalation of volatile residues could be a problem.

Exposure

Routes of exposure could be by (1) consumption of contaminated water and food if such occurred, and (2) inhalation; however, additional data are needed to assess problems caused by volatility.

Chloroform

Data

WS	8,200 ppm	[4]
VP	150.5 torr at 20°C	[4]
Hydrolysis	$t_{1/2}$ 3,500 years	[4]
Light adsorption	180 to 240 nm range	[4]
Pure chloroform is light-sensitive.		[9]

Discussion

- WS indicates leaching and runoff could occur; however, VP is so high that volatility should occur first.
- VP indicates volatility, and light adsorption indicates phototransformation.
- Hydrolysis indicates that the chemical is stable and should not be hydrolyzed in soils or in water. It could be biodegraded or phototransformed if it were to get into the aquatic environment; however, if it were not ingested, there would be no problem. It could come into direct contact with animals through volatilization.

Exposure

(1) Direct contact, (2) ingestion of contaminated water, or (3) inhalation could be the routes of exposure. Chloroform should volatilize and be phototransformed. If it got into the aquatic environment, it should not bioaccumulate, based on WS, and should volatilize from water unless it reached groundwater, where it could be stable. Inhalation should be the most direct route of exposure.

Chloromethane (also Methyl chloride)

Data

WS at 20°C	6,450 to 7,250 mg/L	[4]
Kow	8	[4]
VP at 20°C	3,756 torr	[4]

Discussion

- WS indicates that this chemical should leach, run off, and be biodegraded, and it should not be bioaccumulated and should not be adsorbed.
- Kow indicates that the chemical should not bioaccumulate, and this is in agreement with the value for WS.
- VP indicates that volatility should occur for this chemical; thus inhalation could be a problem. If it were phototransformed, photoproducts and the parent chemical could fall out and contaminate water and food.
- This chemical should be mobile in the environment and could contaminate water and food, and thus could result in contamination of the food chain. Inhalation of volatile residues could be a problem.

Exposure

Routes of exposure could be by (1) consumption of contaminated water and food if such occurred, and (2) inhalation of volatile residues.

Bis (2-Chloromethyl) ether

Data

WS	22,000 mg/L	[4]
VP at 22°C	30 torr	[4]
Kow	2.4	[4]
Hydrolyzes on contact with water	$t_{1/2}$ 10 to 40 sec	[4]

Discussion

- WS indicates that this chemical could run off, leach, and be biodegraded, but it should not bioaccumulate and should not be adsorbed to soil.
- VP indicates volatility; however, this chemical in the presence of moisture should hydrolyze in the atmosphere. Hydrolysis should prevent future exposure of the parent chemical; however, the hydrolytic products need to be studied.

- Kow indicates that this chemical should not bioaccumulate.
- It hydrolyzes with a rapid $t_{1/2}$ of 40 sec. Because the chemical volatilizes and hydrolyzes with moisture, inhalation could be a problem with effects on or in mucous membranes, lungs, and perhaps the digestive system.

Exposure

The main route of exposure could be by inhalation, which could result in problems in the mucous membranes, lungs, and digestive system. Ingestion of contaminated drinking water could be a short-term problem if consumption occurred immediately after contamination; however, this is highly unlikely because of the chemical's rapid hydrolytic rate.

Chloroneb (1,4-Dichloro-2,5-dimethoxybenzene) (also Demosan and Tersan SP)

Data

Pesticide		
WS	8 ppm	[6]
Koc	1,159	[6]

Discussion

- WS indicates that this chemical should be adsorbed to soil, run off with soil, and be bioaccumulated, and it should not leach and should not be biodegraded.
- Koc indicates that this chemical should adsorb in soil and accumulate in the soil environment. Bioaccumulation could be expected, and this chemical could cause food-chain contamination. Residues should be expected in the food chain.

Exposure

Routes of exposure could be by (1) consumption of contaminated water and food if such occurred, and (2) inhalation if the chemical were volatile; however, no data are presented herein on volatility.

2-Chlorophenol (also *o*-Chlorophenol)

Data

Pesticide		
WS at 20°C	28,500 mg/L	[4]
Kow	148	[4]
VP at 20°C	2.2 torr	[4]

Discussion

- WS indicates that this chemical could leach, run off, and be biodegraded, and it should not adsorb in soil and should not be bioaccumulated.
- Kow indicates that bioaccumulation should not occur; however, residues could be expected in the food chain.
- VP indicates volatility, and fallout could contaminate food and water. If the chemical were phototransformed, the transformation products could also contaminate food and water. Inhalation of the parent chemical and its potential photoproducts could be a problem.

Exposure

Routes of exposure could be by (1) ingestion of contaminated food and water and (2) inhalation of a volatile residue.

4-Chlorophenyl phenyl ether

Data

WS	59 mg/L	[4]
Kow	100,000	[4]
VP at 20°C	0.001 torr	[4]

Discussion

- WS indicates that this chemical could bioaccumulate, adsorb to soil, and run off with soil particles, and it should not leach and should not be biodegraded. If Kow were unknown, this prediction would be difficult, as this chemical could go either way.

- Kow indicates bioaccumulation and potential for food-chain contamination.
- VP indicates that volatility may be likely, and phototransformation could occur. Inhalation and fall out of photo products would be a problem.

Exposure

The routes of exposure could be by (1) ingestion of contaminated food or water if such contamination should occur, and (2) inhalation of volatile residues.

Chloroxuron (3-[*p*-(Chlorophenoxy) phenyl]-1,1-dimethlyurea) (also Tenoran)

Data

Pesticide		
WS	0.2 ppm	[6]
Koc	6,790	[6]
Soil TLC-Rf	0.09	[1]

Discussion

- WS indicates that this chemical should adsorb to soil, run off with soil, and be bioaccumulated, and it should not leach and should not be biodegraded.
- Koc indicates that this chemical should be adsorbed to soil, run off with soil, and be bioaccumulated, and it should not be leached and should not be biodegraded.
- Soil TLC indicates that this chemical should not leach and should be adsorbed to soil.
- The three pieces of data complement each other; thus the predictions should be conclusive. The data also indicate that residues could be expected in the food chain, based on the fact that the chemical may be bioaccumulated.

Exposure

Routes of exposure could be by (1) consumption of contaminated water and food if such occurred, and (2) inhalation if the chemical were volatile; however, no data are presented on volatility herein.

Chlorpropham (also Isopropyl N-(3-chlorophenyl) carbamate and CIPC)

Data

WS	88 ppm	[6]
Koc	590	[6]
Soil TLC-Rf	0.18	[1]

Discussion

- WS indicates that this chemical could go either way in relation to leaching, adsorption, runoff, biodegradation, and bioaccumulation.
- Koc indicates that this chemical should not adsorb, and that leaching could occur.
- Soil TLC indicates that this chemical should adsorb to soil to some extent and should not leach.
- The three pieces of data contradict each other; therefore, in this case, the data cannot not be used to make a prediction. Actual laboratory and perhaps field studies will be needed to discern what happens to this chemical in the environment. There will always be chemicals that do not fit the predictive mode, and one most beware of this possibility.

Exposure

Routes of exposure cannot be predicted with the data presented herein. Additional data will be needed to make conclusive predictions.

Chlorthiamid (also 2,5-Dichlorothiobenzamide and Prefix)

Data

Pesticide		
WS	950 ppm	[6]
Koc	107	[6]

Discussion

- WS indicates that this chemical should adsorb to soil, run off with soil, and be bioaccumulated, and it should not leach and should not be biodegraded.
- Koc indicates that this chemical would do just the opposite of what the WS values would predict.
- Additional data would be needed to verify which piece of data could be used in predicting. Even then, actual laboratory and perhaps field data may be needed in order to make a conclusive prediction for this chemical.

Exposure

Routes of exposure cannot be predicted with the data presented above. Additional data will be needed for predictions.

Chrysene (also 1,2-Benzphenanthrene)

Data

WS at 25°C	0.002 mg/L	[4]
Kow	407,380	[4]
VP at 20°C	6.3×10^{-7} torr	[4]

Discussion

- WS indicates that this chemical could adsorb in soil, run off with soil, and be bioaccumulated, and it should not leach and should not be biodegraded.
- Kow indicates bioaccumulation, which could cause food-chain contamination. Residues could be expected in the food chain.
- VP indicates that volatility is of no concern.

Exposure

Routes of exposure could be by ingestion of contaminated food and water if such contamination should occur.

Crufomate (4-*tert*-Butyl-2-chlorophenyl methyl methylphosphoramidate) (also Ruelene)

Data

Pesticide		
WS	200 ppm	[6]
Kow	2,780	[6]

Discussion

- WS indicates that this chemical could go either way in relation to adsorption, leaching, runoff, biodegradation, and bioaccumulation.
- Kow indicates that this chemical should bioaccumulate and could cause food-chain contamination. Residues could be expected in the food chain.
- The data complement each other; thus this chemical should be expected to accumulate in the environment, and residues could be expected in the food chain.

Exposure

Routes of exposure could be by (1) consumption of contaminated water and food if such occurred, and (2) inhalation if the chemical were volatile; however, no data are presented herein on volatility.

Cyanazine (also 2-[[4-Chloro-6-(ethylamino)-*s*-triazin-2-yl] amino]-2-methylpropionitrile), Bladex and Payze)

Data

Pesticide		
WS	171 ppm	[6]
Koc	200	[6]
Kow	150	[6]
BCF—static water	0	[6]

Discussion

- WS indicates that this chemical could go either way in relation to leaching, runoff, adsorption, biodegradation, and bioaccumulation.
- Koc indicates that the chemical should not adsorb.
- Kow indicates that the chemical should not bioaccumulate.
- BCF indicates that the chemical should not bioaccumulate and could be metabolized in animals.
- If this chemical is released into the environment, it could be mobile, could be biodegraded, and should not bioaccumulate. There is a possibility that residues could be present in the food chain; however, there are no data presented on biodegradation and metabolism that would allow one to assess dissipation.

Exposure

Routes of exposure could be by (1) consumption of contaminated water and food if such occurred, and (2) inhalation if the chemical were volatile; however, no data are presented herein on volatility. Additional data would be needed to discern whether biodegradation and metabolism could prevent residues from being of concern.

Cycloate (also *S*-Ethyl cyclohexylethylthiocarbamate, Eurex, and Ro-Neet)

Data

Pesticide		
WS	85 ppm	[6]
Koc	345	[6]

Discussion

- WS indicates that this chemical could go either way in regard to leaching, runoff, adsorption, biodegradation, and bioaccumulation.
- Koc indicates that the chemical could leach, run off, and be biodegraded, and it should not bioaccumulate and should not be adsorbed in soil.

- It is predicted that this chemical could be mobile and could contaminate the food chain if it were not biodegraded or metabolized rapidly enough. This prediction is based on the value for Koc and somewhat on the value for WS. Additional data are needed for a more precise prediction.

Exposure

Routes of exposure could be by (1) consumption of contaminated water and food if such occurred, and (2) inhalation if the chemical were volatile; however, no data are presented herein on volatility.

2,4-D (or 2,4-Dichlorophenoxyacetic acid)

Data

Pesticide		
WS	900 ppm	[6]
Koc	20	[6]
Kow	37	[6]
BCF—static water	0	[6]
Soil TLC-Rf	0.5 to 0.69; class 4	[5]
Protein-bound		[10]

Discussion

- WS indicates that the chemical should not bioaccumulate, and that it could leach and run off.
- Koc indicates that soil adsorption should not occur, and this is supported by WS and soil TLC.
- Kow indicates that the chemical should not bioaccumulate, and food-chain contamination should not occur. This is further supported by a BCF of zero, which shows no bioaccumulation and possible metabolism in animals.
- Soil TLC indicates that leaching could occur, and this is supported by WS. The chemical could leach or run off to water and be stable in water unless degraded by microbes.

Exposure

2,4-D could leach and run off, causing contamination to the aquatic environment. Exposure could be by ingestion of contaminated water. Food-chain contamination is unlikely because of the zero BCF, which indicates metabolism by animals.

It should be noted that Yip and Ney [10] reported that 2,4-D take-up by plants and ingestion by cows resulted in residues in cow's milk, which were bound in the milk protein. The WS did not indicate this, nor could it be predicted. This is a good example of why a battery of testing and toxicological studies is needed. Since the chemical got into cow's milk, it could also get into human milk.

Dalapon (2,2-Dichloropropionic acid)

Data

Pesticide		
WS	502,000 ppm	[6]
Kow	6	[6]
BCF—static water	3	[6]
Soil TLC-Rf	0.96	[1]

Discussion

- WS indicates that the chemical could leach, run off, and be biodegraded, and it should not adsorb to soil and should not be bioaccumulated.
- Kow indicates that it should not bioaccumulate.
- BCF indicates that it should not bioaccumulate, but residues may be found in the food chain.
- Soil TLC-Rf indicates that it could leach in soils.

Exposure

Routes of exposure could be by (1) drinking contaminated water, (2) ingestion of contaminated food, and (3) inhalation if the chemical were volatile; however, no data on volatility are presented herein.

DBCP (also 1,2-Dibromo-3-chloropropane, Nemagon, and Fumazone)

Data

Pesticide		
WS	1,230 ppm	[6]
Koc	129	[6]
VP at 21°C	0.8 mm Hg	[8]

Discussion

- WS indicates that this chemical should leach, run off, and be biodegraded, and it should not be adsorbed and should not be bioaccumulated. This is supported by the value for Koc.
- Koc indicates that this chemical should not be adsorbed and should not be bioaccumulated, and it should leach, run off, and be biodegraded.
- VP indicates that this chemical is volatile; thus inhalation could be a problem. Volatile residues and—if the chemical were phototransformed—both parent and phototransformation products could fall out and contaminate water and the food chain.
- This chemical, if not biodegraded rapidly, could cause contamination of water and food. Inhalation would be a problem, based on its volatility.

Exposure

Routes of exposure could be by (1) inhalation of volatile residues, as well as, perhaps, phototransformation products, and (2) consumption of contaminated water and food if such occurred; however, biodegradation may be rapid enough to prevent exposure, but no data are presented on biodegradation herein.

DDD (Dichloro diphenyl dichloroethane)

Data

Pesticide		
WS	0.005 ppm	[6]
Kow	1,047,000	[6]
BCF—static water	63,830	[6]

Discussion

- WS indicates that this chemical should adsorb to soil, run off with soil, and bioaccumulate, and it should not be leached and should not be biodegraded.
- Kow indicates that the chemical should bioaccumulate and could cause contamination to the food chain. This prediction is supported by the value given for BCF and WS.
- BCF indicates that this chemical should bioaccumulate and should cause food-chain contamination. It should not be metabolized to any extent in animals.
- The chemical should contaminate the food-chain if released into the environment. Residues should be expected in the food chain, and very little metabolism is expected in animals.

Exposure

Routes of exposure (1) should be by consumption of contaminated water and food if the chemical were released into the environment, and (2) could be by inhalation if it were volatile; however, no data are presented herein on volatility.

DDE

Data

Pesticide		
WS	0.01 ppm	[6]
Kow	583,000	[6]
BCF—static water	27,400	[6]

Discussion

- WS indicates that this chemical should adsorb to soil, run off with soil, and be bioaccumulated, and it should not be leached and should not be biodegraded.
- Kow indicates that the chemical should bioaccumulate and could cause contamination to the food chain.
- BCF indicates that the chemical should bioaccumulate in the food chain, and little metabolism is expected in animals.
- This chemical should bioaccumuate and should cause contami-

nation to the food chain. Very little metabolism is expected in animals. Residues should be expected in the food chain.

Exposure

Routes of exposure (1) should be by consumption of contaminated water and food if such occurred, and (2) could be by inhalation if the chemical were volatile; however, no data are presented herein on volatility.

DDT (Dichloro diphenyl trichloroethane)

Data

Pesticide		
WS	0.0017 ppm	[6]
Koc	238,000	[6]
Kow	960,00	[6]
BCF—flowing water	61,600	[6]
BCF—static water	84,500	[6]

Discussion

- WS indicates that this chemical should adsorb to soil, run off with soil, and bioaccumulate, and it should not be leached and should not be biodegraded.
- Koc indicates that this chemical should adsorb to soil.
- Kow indicates that this chemical should bioaccumulate and could cause food-chain contamination, and this is supported by the BCF data.
- BCF indicates that this chemical should bioaccumulate in the food chain; thus residues could be expected.
- This chemical should be persistent in all environmental compartments, and residues could be expected throughout the food chain.

Exposure

Routes of exposure could be by (1) consumption of contaminated food and water if the chemical were released into the environment, and (2) inhalation if such occurred; however, no data are presented herein on volatility.

Dialifor (O,O-Diethyl S-(2-chloro-1-phthalimidoethyl) phosphorodithioate) (also dialifos and Torak)

Data

Pesticide		
WS	0.18 ppm	[6]
Kow	49,300	[6]

Discussion

- WS indicates that this chemical should not leach and should not be biodegraded, and that it should be adsorbed and be bioaccumulated, and it can run off with soil.
- Kow indicates that this chemical should bioaccumulate in animals and could cause food-chain contamination. Residues should be expected in the food chain.
- Both pieces of data indicate bioaccumulation and the potential for food-chain contamination. Biodegradation is not expected.

Exposure

Routes of exposure could be by (1) consumption of contaminated water and food if such occurred, and (2) inhalation if the chemical were volatile; however, no data are presented herein on volatility.

Diallate (also *S*-(2,3-Dichloroallyl) diisopropylthiocarbamate, Avadex, and DATC)

Data

Pesticide		
WS	14 ppm	[6]
Koc	1,900	[6]

Discussion

- WS indicates that this chemical should be adsorbed in soil and be bioaccumulated, and it can run off with soil; also it should not be biodegraded and should not be leached.

- Koc indicates that this chemical should adsorb in soil and could cause food-chain contamination. If this occurred, then residues should be expected in the food chain.
- This chemical could result in food-chain contamination and residues in food sources. The chemical is a carbamate derivative; therefore, it has the potential to be metabolized in animals.

Exposure

Routes of exposure could be by (1) consumption of contaminated water and food if such occurred, and (2) inhalation if the chemical were volatile; however, no data are presented herein on volatility.

Diamidofos (Phenyl *N,N'*-dimethylphosphorodiamidate) (also Nellite and Dowco 169)

Data

Pesticide		
WS	50,000 ppm	[6]
Koc	32	[6]
BCF—flowing water	1	[6]

Discussion

- WS indicates that this chemical should leach, run off, and be biodegraded, and it should not be adsorbed and should not be bioaccumulated. This prediction is supported by the values for Koc and BCF.
- Koc indicates that the chemical should not be adsorbed in soil.
- BCF indicates that the chemical should not bioaccumulate and could be metabolized in animals.
- If this chemical were released into the environment, it could contaminate water and the food chain; however, it should be biodegraded in soil and metabolized in animals. There is a possibility that residues could be found in the food chain, but they should dissipate in time.

Exposure

Routes of exposure could be by (1) consumption of contaminated water and food if such occurred; however, if dissipation is rapid, these residues may not be a problem, and additional data should support this prediction. Also exposure could be by (2) inhalation if the chemical were volatile; however, no data are presented herein on volatility.

Diazinon (also O,O-Diethyl O-(2-isopropyl-6-methyl-4-pyrimidinyl) phosphorothioate and Spectracide)

Data

Pesticide		
WS	40 ppm	[6]
BCF—flowing water	35	[6]

Discussion

- WS indicates that this chemical could go either way in relation to leaching, runoff, adsorption, biodegradation, and bioaccumulation.
- BCF indicates the possibility of a residue, but not bioaccumulation. This chemical is an organo phosphate and is likely to cause death prior to bioaccumulation unless it is metabolized in animals. Residues could be expected in animals if the food chain is contaminated.
- Additional data would be needed to discern or predict what may happen to this chemical in soil, because the WS value in this case is not sufficient for us to make conclusive predictions.

Exposure

Routes of exposure could be by (1) drinking contaminated water, (2) ingestion of contaminated food, and (3) inhalation if the chemical were volatile; however, no data are presented herein on volatility.

Dibromochloromethane

Data

Kow	123	[4]
VP at 10.5°C	15 torr	[4]

Discussion

- Kow indicates that this chemical should not bioaccumulate. Based on the Kow value, it is predicted that this chemical could leach, run off, and be biodegraded, and it should not adsorb to soil.
- VP indicates volatility; thus inhalation could be a problem. If the chemical were phototransformed, the photoproducts and the parent chemical could fall out and contaminate water and food, thus resulting in contamination of the food chain.
- This chemical could be mobile and could contaminate water and food. Inhalation of volatile residues could be a problem. Fallout of the parent chemical and photoproducts could cause contamination of the food chain and inhalation problems.

Exposure

Routes of exposure could be by (1) consumption of contaminated water and food if such occurred, and (2) inhalation of volatile residues.

Dicamba (3,6-Dichloro-*o*-anisic acid) (also Banvel D)

Data

Pesticide

WS	4,500 ppm	[6]
Koc	0.4	[6]
Soil TLC-Rf	0.96	[1]

Discussion

- WS indicates that this chemical should leach, run off, and be biodegraded, and it should not be adsorbed to soil and should not be bioaccumulated.

- Koc indicates that this chemical should not be adsorbed to soil and should not be bioaccumulated; thus it should be biodegraded in soils.
- Soil TLC indicates that the chemical should leach and could contaminate the aquatic environment.
- This chemical should leach and could contaminate the aquatic environment; however, it may be biodegraded prior to reaching groundwater, but no data are presented on biodegradation to support this prediction. The WS value is indicative of biodegradation, but the soil TLC is indicative of leaching. Hopefully one effect will offset the other. Additional data would help us to predict biodegradation. Residues may be expected in the food chain if biodegradation is not rapid.

Exposure

Routes of exposure could be by (1) consumption of contaminated water and food if such occurred, and (2) inhalation if the chemical were volatile; however, no data are presented herein on volatility.

Dichlobenil (2,6-Dichlorobenzonitrile) (also Casoron)

Data

Pesticide		
WS	0.18 ppm	[6]
Koc	235	[6]
BCF—static water	55	[6]
Soil TLC-Rf	0.22	[1]

Discussion

- WS indicates that this chemical should adsorb to soil (this does not agree with Koc), be bioaccumulated (this does not agree with BCF), and run off with soil, and it should not leach and should not be biodegraded.
- Koc indicates that this chemical should not adsorb to soil and should leach (this is not in agreement with soil TLC).
- BCF indicates that this chemical should not bioaccumulate, and this is not in agreement with WS. BCF indicates that the chemi-

cal could be metabolized in animals; however, no data are presented on animal metabolism. The chemical, if released into the environment, could cause residues in the food chain.

- Soil TLC indicates that the chemical adsorbs to soil, which is not in agreement with Koc but agrees with WS.
- WS and soil TLC are in agreement, in that the chemical should adsorb to soil; however, Koc does not agree. BCF indicates that residues could be expected in the food chain, and they may be slowly metabolized. BCF indicates that bioaccumulation is unlikely, and this is not in agreement with WS and soil TLC. The bottom line is that residues could be expected in the food chain. Additional data would help clarify the difference in data; however, exposure can be predicted.

Exposure

Routes of exposure could be by (1) consumption of contaminated water and food if such occurred, and (2) inhalation if the chemical were volatile; however, no data are presented herein on volatility. Additional data would help us to make a more conclusive prediction.

Dichlofenthion (O-(2,6-Dichlorophenyl) O,O-diethyl phosphorothioate) (also VC-13 and Nemacide)

Data

Pesticide		
WS	0.245 ppm	[6]
Kow	137,000	[6]

Discussion

- WS indicates that this chemical should adsorb to soil, run off with soil, and be bioaccumulated, and it should not be biodegraded and should not be leached. The value for Kow supports this prediction.
- Kow indicates that the chemical should bioaccumulate, and this could cause food-chain contamination
- If this chemical were released into the environment, it could cause contamination of the food-chain.

Exposure

Routes of exposure could be by (1) consumption of contaminated water and food is such occurred, and (2) inhalation if the chemical were volatile; however, no data are presented herein on volatility.

o-Dichlorobenzene (also 1,2-Dichlorobenzene)

Data

WS at 25°C	145 mg/L	[4]
Kow	2,399	[4]
VP at 25°C	1.5 torr	[4]

Discussion

- WS in conjunction with Kow indicates that this chemical could adsorb in soil, run off with soil, and be bioaccumulated, and it should not leach and should not be biodegraded.
- Kow indicates bioaccumulation and food-chain contamination. Residues could be expected in the food chain.
- VP indicates volatility; fallout could contaminate water and food. If the chemical were phototransformed, transformation products could also fall out to contaminate food and water. Inhalation of the parent compound and its potential photoproducts could be a problem.

Exposure

Routes of exposure could be by (1) ingestion of contaminated food and water and (2) inhalation of volatile residues.

p-Dichlorobenzene

Data

Pesticide—mothballs		[6]
WS	79 ppm	[6]
Kow	2,455	[6]
VP at 20°C	0.6 torr	[4]
VP at 25°C, calculated	1.18 torr	[4]
BCF—flowing water	215	[6]

Discussion

- WS indicates that the chemical is soluble and could go either way—be mobile or nonmobile.
- Kow indicates that bioaccumulation and food-chain contamination could occur, and this is supported by a BCF of 215 in fish. Because the Kow is high, we can predict that this chemical should not leach.
- VP indicates volatility, and inhalation could be a problem.
- The chemical will cause food-chain contamination if it gets to the aquatic environment. It is volatile enough to volatilize from surfaces and form surface waters. Drinking water, if contaminated, could be a problem.

Exposure

The main route of exposure should be inhalation. If phototransformation did not occur (no data are given here), then fallout to noncontaminated areas could be a problem.

4,4′-Dichlorobiphenyl

Data

WS	0.062 ppm	[6]
Kow	380,000	[6]
BCF—flowing water	215	[6]

Discussion

- WS indicates that this chemical should adsorb to soil, run off with soil, and be bioaccumulated, and it should not leach and should not be biodegraded. The Kow value supports this prediction; however, the BCF value does not. Perhaps the BCF value is an indication that the chemical is being metabolized in animals.
- Kow indicates that the chemical is lipo-soluble; thus it should bioaccumulate in animals. BCF does not support this prediction, but WS does.
- BCF indicates that the chemical should result in residues in the food chain; however, bioaccumulation should not occur. Metabolism cannot be predicted without additional data.

- This chemical should adsorb to soil and could contaminate the food chain, and thus could cause residues in food and water.

Exposure

Routes of exposure could be by (1) consumption of contaminated water and food if such occurred, and (2) inhalation if the chemical were volatile; however, no data are presented herein on volatility.

Dichlorodifluoromethane (also Freon-12 and Fluorocarbon-12)

Data

WS	280 mg/L	[4]
Kow	144.54	[4]
VP at 20°C	4,306 torr	[4]

Discussion

- WS indicates that this chemical could go either way; however, Kow indicates that bioaccumulation should not occur; therefore, it could leach, run off and biodegrade, and should not adsorb to soil and not be bioaccumulated.
- Kow indicates that bioaccumulation should not occur.
- VP indicates volatility, and this could cause inhalation problems. If the chemical were phototransformed, photoproducts and the parent chemical could fall out and contaminate the environment.
- This chemical is volatile and could present inhalation problems with the parent chemical and potential photoproducts. This type of chemical is known to damage the ozone layer in the atmosphere. It is unlikely that residues could be found in food, and it should not last long in water because of its volatility.

Exposure

The route of exposure could be inhalation of volatile residues. The major concern with this chemical should be the degradation of the ozone layer.

2,4-Dichlorophenol

Data

WS at 20°C	4,500 mg/L	[4]
Kow	562	[4]
VP at 20°C	0.12 torr	[4]

Discussion

- WS indicates that this chemical could leach, run off, and be biodegraded, and it should not adsorb to soil and should not be bioaccumulated.
- Kow indicates the possibility of bioaccumulation. Residues could be expected in the food chain.
- VP indicates volatility; fallout could cause food-chain contamination. If the chemical were phototransformed, transformation products also could fall out and cause food-chain contamination. Inhalation of the parent chemical and its photoproducts could be a problem.

Exposure

Routes of exposure could be by (1) ingestion of contaminated food and water and (2) inhalation of volatile residues.

3,6-Dichloropicolinic acid

Data

Pesticide		
WS	1,000 ppm	[6]
Koc	2	[6]

Discussion

- WS indicates that this chemical should leach, run off, and be biodegraded, and it should not adsorb in soil and should not be bioaccumulated.
- Koc indicates that this chemical should not adsorb to soil and should not be bioaccumulated. WS and Koc are in agreement.

- There are insufficient data to predict whether residues could occur in the food chain. If the chemical reached the aquatic environment, then residues could be found in the food chain unless it were rapidly metabolized. This chemical could leach and run off into the aquatic environment.

Exposure

Routes of exposure could be by (1) consumption of contaminated water and food and (2) inhalation if the chemical were volatile; however, no data are presented herein on volatility.

Dichlorovos (also 2,2-Dichlorovinyl dimethyl phosphate, DDVP, and Vapona)

Data

Pesticide		
WS	10,000 ppm	[6]
Kow	25	[6]

Discussion

- WS indicates that this chemical could leach, run off, and be biodegraded, and it should not adsorb to soil and should not be bioaccumulated.
- Kow indicates that bioaccumulation should not occur.

Exposure

Routes of exposure could be by (1) ingestion of contaminated food and water if such occurred, and (2) inhalation if the chemical were volatile; however, no data on volatility are presented herein.

Dieldrin (Hexachloroepoxyoctahydro-endo, exo-dimethanonaphthalene 85% and related compounds 5%)

Data

Pesticide		
WS	0.022 ppm	[6]

BCF—flowing water	5,800	[6]
BCF—static water	4,420	[6]

Discussion

- WS indicates that this chemical could be adsorbed in soil, run off with soil, and be bioaccumulated, and it should not leach and should not be biodegraded.
- BCF indicates that bioaccumulation and food-chain contamination could occur. Residues could be expected in the food chain.

Exposure

Routes of exposure could be by (1) ingestion of contaminated food and water and (2) inhalation if the chemical were volatile; however, no data are presented herein on volatility. Food-chain contamination is highly likely, and so are residues of this chemical in food, based on the BCF values given.

Diethylaniline

Data

WS	670 ppm	[6]
Kow	9	[6]
BCF—static water	120	[6]

Discussion

- WS indicates that this chemical could go either way in relation to leaching, runoff, adsorption, biodegradation, and bioaccumulation. The use of Kow and BCF values in this case supports the prediction that the chemical could leach, run off, and be biodegraded, and that it should not be bioaccumulated and should not be adsorbed.
- Kow indicates that this chemical should not adsorb to soil and should not be bioaccumulated. This prediction is in agreement with WS.
- BCF indicates that the chemical should not accumulate in animals; however, residues could be expected in the food chain

if contamination occurred. It is possible that metabolism could occur in animals.

- This chemical, if released into the environment, could contaminate water and the food chain. If this occurred, then residues could be expected in the food chain.

Exposure

Routes of exposure could be by (1) consumption of contaminated water and food if such occurred, and (2) inhalation if the chemical were volatile; however, no data are presented herein on volatility.

Di-2-ethylhexyl phthalate (also DEHP)

Data

Pesticide		
WS	0.6 ppm	[6]
Kow	9,500	[6]
BCF—flowing water	380	[6]
BCF—static water	130	[6]

Discussion

- WS indicates that this chemical should be adsorbed in soil and be bioaccumulated and could run off with soil particles, and that it should not leach and should not be biodegraded.
- Kow indicates that this chemical could bioaccumulate.
- BCF indicates that the chemical may not bioaccumulate, perhaps because it is metabolized. The data do indicate that a chemical residue could be expected in the food chain. There is a contradiction between WS and Kow values and the BCF; however, that is acceptable because residues of the chemical can be expected in the food chain although bioaccumulation may not occur. Residues could occur in the food chain if the chemical contaminated water.

Exposure

Routes of exposure could be by (1) ingestion of contaminated food, (2) drinking contaminated water, and (3) inhalation if the chemical were volatile; however, no data on volatility are presented herein.

Diethyl phthalate (also DEP)

Data

WS	1,000 mg/L	[4]
VP	0.05 torr	[4]
Kow	26,303	[4]

Discussion

- WS indicates that the chemical could leach, run off, and be biodegraded, and that it should not bioaccumulate and should not be adsorbed in soil.
- VP indicates that the chemical could be volatile, and could fall out and contaminate food and water. If it were phototransformed, the transformation products also could fall out and contaminate food and water. Inhalation of the parent chemical and its photoproducts could be a problem.
- Kow indicates that bioaccumulation and food-chain contamination are likely. This prediction contradicts WS; however, it appears to be a valid prediction with no reason to alter it unless additional data are presented to justify a change.

Exposure

Routes of exposure could be by (1) inhalation of volatile residues and (2) ingestion of contaminated food and water.

Dimethoate (*O,O*-Dimethyl *S*-[(methylcarbanoyl) methyl]phosphorodithioate) (also Cygon)

Data

Pesticide		
WS	25,000 ppm	[6]
Kow	0.51	[6]

Discussion

- WS indicates that this chemical could leach, run off, and be biodegraded, and it should not bioaccumulate and should not be adsorbed in soil.
- Kow indicates that the chemical should not bioaccumulate; however, residues of the chemical may be found in the food chain.

Exposure

Routes of exposure could be by (1) ingestion of contaminated water or food if such occurred, and (2) inhalation if the chemical were volatile; however, no data are presented herein on volatility.

Dimethylnitrosamine

Data

WS	miscible	[4]
Kow	1.15	[4]
Adsorbs light	up to 400 nm	[4]
Photolysis in atmosphere	$t_{1/2}$ is 1 hour	[4]

Discussion

- WS indicates that this chemical could leach, run off, and be biodegraded, and that it should not bioaccumulate and should not be adsorbed in soil.
- Kow indicates that bioaccumulation should not occur.
- If volatile or on surfaces, this chemical could be phototransformed to other chemicals in hours. Nitrosamines are known to cause cancer; therefore, all phototransformation products are of concern. Fallout of this chemical and its photoproducts could contaminate water and the food chain. Inhalation, if the chemical is found to be volatile, is of concern for the parent chemical and its photoproducts.

Exposure

Routes of exposure (1) could be by ingestion of contaminated food or water, and (2) inhalation could be a problem, if one were continuously exposed to a volatile residue, if the chemical were found to be volatile. If volatile, the chemical should be phototransformed in the atmosphere. Phototransformation products may present an exposure problem if the formation of nitrosamines occurs.

Dimethyl phthalate (also DMP)

Data

WS	4,000 mg/L	[4]
VP at 20°C	<0.01 torr	[4]
Kow	2,630	[4]

Discussion

- WS indicates that this chemical could leach, run off, and be biodegraded, and it should not bioaccumulate and should not be adsorbed in soil.
- VP indicates that this chemical could be volatile. If it were volatile, phototransformation products could be formed and fall out to contaminate water and the food chain. Inhalation of the parent chemical and its potential photoproducts could be a problem.
- Kow indicates that bioaccumulation is likely, as is food-chain contamination. Kow contradicts WS for bioaccumulation; however, bioaccumulation is expected. A residue of DMP could be expected in water and the food chain.

Exposure

Routes of exposure could be by (1) ingestion of contaminated water and food and (2) inhalation of volatile DMP.

Dimilin (also Diflubenzuron (N-[[(4-Chlorophenyl) amino] carbonyl]-2,6-difluorobenzamide))

Data

Pesticide		
WS	0.2 ppm	(6)
Koc	6,790	(6)

Discussion

- Looking at the chemical structure of Dimilin, an experienced chemist could predict some of the following:
  a. Breakdown to *p*-Chloroaníline, a pesticide, which is practically insoluble in water [8]. It is possible that this breakdown product could form azo compounds, which are known carcinogens.
  b. Breakdown to difluorobenzoic acid, with WS of 1 gram in 2.3 ml of water, its solubility increasing with alkaline conditions [8].
- The breakdown products given above are only an assumption, but a good one. There could be many more breakdown products. Such assumptions cannot predict the cause of breakdown, but biodegradation and metabolism would be a good guess for it.
- WS indicates that leaching should not occur, adsorption to soil could occur, runoff with soil particles is possible, and bioaccumulation should occur unless the chemical is metabolized or biodegraded.
- Koc indicates that this chemical should adsorb to soil, and this is supported by the WS.
- Under alkaline conditions this chemical could form water-soluble chemicals, and these products could leach and contaminate sources of water.

Exposure

Exposure could be through bioaccumulation and food chain contamination. Inhalation could be a problem if the chemical were encountered in areas of aircraft applications.

Dinitramine (N^3,N^3-Diethyl-2,4-dinitro-6-(trifluoromethyl)-1,3-benzenediamine) (also Cobex)

Data

Pesticide

WS	<1 ppm	[6]
Koc	3,600	[6]

Discussion

- WS indicates that this chemical should be adsorbed to soil, should be bioaccumulated, and can run off with soil, and it should not be biodegraded and should not be leached.
- Koc indicates that the chemical should bioaccumulate and could cause contamination in the food chain. Residues could be expected in the food chain.
- This chemical, if released into the environment, could bioaccumulate and cause residues in the food chain.

Exposure

Routes of exposure could be by (1) consumption of contaminated water and food if such occurred, and (2) inhalation if the chemical were volatile; however, no data on volatility are presented herein.

4,6-Dinitro-*o*-cresol

Data

Kow	708	[4]

Discussion

- Kow indicates potential for bioaccumulation, and that residues could be expected in the food chain if contamination occurred. Kow also indicates that WS could be the cause of leaching, runoff, and biodegradation, but should not be the cause of adsorption in soil or bioaccumulation.

Exposure

Routes of exposure could be by (1) ingestion of contaminated food and water if such occurred, and (2) inhalation if the chemical were volatile; however, no date are presented herein to indicate volatility.

2,4-Dinitrophenol

Data

WS at 18°C	5,600 mg/L	[4]
Kow	34	[4]

Discussion

- WS indicates that this chemical could leach, run off, and be biodegraded, and it should not adsorb in soil and should not be bioaccumulated.
- Kow indicates that bioaccumulation should be no problem.

Exposure

Routes of exposure could be by (1) ingestion of contaminated food and water and (2) inhalation if the chemical were found to be volatile; however, no data are presented herein on volatility.

2,4-Dinitrotoluene

Data

WS at 22°C	270 mg/L	[4]
Kow	102	[4]
VP at 59°C	0.0013 torr	[4]

Discussion

- WS indicates that this chemical could leach, run off, and be biodegraded, and it should not bioaccumulate and should not be adsorbed in soil.
- Kow indicates that this chemical should not bioaccumulate, and this supports the predictions based on WS.

- VP indicates potential for volatility; fallout could contaminate food and water. If the chemical were phototransformed, transformation products could fall out, contaminating food and water. Inhalation of the parent compound and its potential photoproducts could be a problem.

Exposure

Routes of exposure could be by (1) ingestion of contaminated food and water and (2) inhalation of volatile residues.

Dinoseb (2-*sec*-Butyl-4,6-dinitrophenol) (also DNBP and Dyanap)

Data

Pesticide		
WS	50	[6]
Koc	4	[6]
Kow	4,900	[6]

Discussion

- WS indicates that this chemical could go either way in regard to leaching, runoff, adsorption, biodegradation, and bioaccumulation.
- Koc indicates that the chemical should not adsorb in soil.
- Kow indicates that the chemical should bioaccumulate.
- The three pieces of data differ, so that it is difficult to make a prediction in regard to leaching, adsorption, runoff, and biodegradation. Based on the value given for Kow, one could predict that residues should be expected in water and food if this chemical were released into the environment. There are no data to assess how long the residues would remain in water and food. Additional data would be needed on biodegradation and metabolism in animals to discern dissipation.

Exposure

Routes of exposure could be by (1) consumption of contaminated water and food if such occurred, and (2) inhalation if the chemical were volatile; however, no data are presented herein on volatility.

Diphenylnitrosamine

Data

Kow	372	[4]

Discussion

- Kow indicates that bioaccumulation should not occur. Kow also indicates that the WS could be in a range that could cause leaching, runoff, and biodegradation, and should not cause adsorption in soil and bioaccumulation.

Exposure

Routes of exposure could be by (1) ingestion of contaminated food and water if such occurred, and (2) inhalation if the chemical were volatile; however, no data on volatility are presented herein.

Diphenyl oxide (also Phenyl ether)

Data

WS	21 ppm	[6]
Kow	15,800	[6]
BCF—flowing water	196	[6]

Discussion

- WS indicates that this chemical could go either way in regard to leaching, runoff, adsorption, bioaccumulation, and biodegradation.
- Kow indicates that the chemical should bioaccumulate and could contaminate the food chain.
- BCF indicates that the chemical should not bioaccumulate, and this is in disagreement with the data given for Kow. However, it may not be in disagreement if the chemical is metabolized in animals. Additional data would be needed to discern metabolism; however, the prediction is that the chemical could adsorb in soil and could be metabolized in animals.
- This chemical, if released into the environment, could contami-

nate the food chain; however, the residues could be metabolized in animals. Residues should be expected in animals, but should dissipate over time.

Exposure

Route of exposure could be by (1) consumption of contaminated water and food if such occurred, and (2) inhalation if the chemical were volatile; however, no data are presented herein on volatility.

Di-*n*-propylnitrosamine

Data

WS at 25°C	9,895 mg/L	[4]
Kow	20	[4]

Discussion

- WS indicates that this chemical could leach, run off, and be biodegraded, and it should not bioaccumulate and should not be adsorbed in soil.
- Kow indicates that bioaccumulation should not occur.

Exposure

Routes of exposure could be by (1) ingestion of contaminated food and water if that occurred, and (2) inhalation if the chemical were found to be volatile; however, no data are presented herein on volatility.

Disulfoton (also *O,O*-Diethyl *S*-[2-(ethylthio) ethyl] phosphorodithioiate, Di-Syston, Dithiodemeton, Thiodemeton, etc.)

Data

Pesticide

WS	25 ppm	[6]
Koc	1,780	[6]

Discussion

- WS indicates that this chemical should adsorb to soil, run off with soil, and be bioaccumulated, and it should not be biodegraded and should not be leached.
- Koc indicates that the chemical should bioaccumulate, and this is supported by the value for WS.
- This chemical should adsorb to soil and could bioaccumulate in the food chain. If released into the environment, it could cause residues in the food chain.

Exposure

Routes of exposure could be by (1) consumption of contaminated food if such occurred, and (2) inhalation if the chemical were volatile; however, no data are presented herein on volatility.

Diuron (3-(3,4-Dichlorophenyl)-1,1-dimethylurea)

Data

WS	42 ppm	[6]
Koc	400	[6]
Kow	94	[6]
Soil TLC-Rf	0.24	[1]

Discussion

- WS indicates the chemical could go either way with respect to leaching, runoff, biodegradation, adsorption in soil, and bioaccumulation. WS alone could not be used to predict the above; additional data would be needed.
- Koc indicates that adsorption should not occur.
- Kow indicates that bioaccumulation should not occur.
- Soil TLC indicates that leaching should not occur.
- One piece of data does not support the other. For this chemical, actual tests would be needed to discern leaching, adsorption, runoff, biodegradation, and bioaccumulation. The data presented indicate that the chemical could result in chemical residues in the food chain but not be bioaccumulated.

Exposure

Routes of exposure could be by (1) ingestion of contaminated water or food if such occurred, and (2) inhalation if the chemical were volatile; however, no data are presented on volatility herein.

DSMA (also Disodium methanearsonate)

Data

Pesticide

WS	254,000 ppm	[6]
Koc	770	[6]

Discussion

- WS indicates that this chemical could leach, run off, and be biodegraded, and it should not adsorb to soil and should not be bioaccumulated.
- Koc indicates that this chemical should not adsorb in soil. Adsorption could be expected for arsenical compounds; however, only the experienced person would know this.
- Arsenical compounds have been found to contaminate the environment. Additional data would be needed for a more inclusive prediction.

Exposure

Routes of exposure could be by (1) ingestion of contaminated food and water if such occurred, and (2) inhalation if the chemical were volatile; however, no data are presented on volatility herein.

Dursban (also Chlorpyrifos (O,O-Diethyl *O*-(3,5,6-trichloro-2-pyridyl phosphorothioate))

Data

Pesticide—termite control

WS	0.3 ppm	[6]
Koc	13,600	[6]

Kow	97,700	[6]
BCF—static water	320	[6]
BCF—flowing water	450	[6]

Discussion

- WS indicates that the chemical should not leach or run off with water. Runoff with soil particles is possible.
- Koc indicates adsorption to soil, which should prevent leaching.
- Kow indicates bioaccumulation in animals.
- BCF indicates bioaccumlation and food-chain contamination.
- Based on the data, the chemical should not leach and contaminate aquatic environments. Food-chain contamination is possible if it is taken up by plants, and the plants are eaten by animals.

Exposure

It is possible that the chemical could be taken up by plants grown in soils containing it; so food-chain contamination is possible. Dursban is an organic phosphate, and its toxicity could be a problem to animals eating food or drinking water containing its residue. Inhalation could be a problem if it were volatile; however, no data are presented herein on volatility.

Endothall (7-Oxabicyclo (2.2.1) heptane-2,3-dicarboxylic acid)

Data

Pesticide

WS	100,000 ppm	[6]
BCF—static water	0	[6]

Discussion

- WS indicates that this chemical should leach, run off, and be biodegraded, and it should not be bioaccumulated and should not be adsorbed.
- BCF indicates the chemical should not bioaccumulate, and this is supported by the value for WS.

- The value for BCF indicates that the chemical may be metabolized in animals. Short-term residues may be expected in the food chain if the chemical is released into the environment.

Exposure

Routes of exposure could be by: (1) consumption of contaminated water and food, but the exposure could be short-term if metabolism and biodegradation were rapid; and (2) inhalation if the chemical were volatile, but no data are presented herein on volatility.

Endrin (Hexachloroepoxyoctahydro-endo, endo-dimethanonaphthalene (also 1,2,3,4,10,10-hexachloro-6,7-epoxy-1,4,4a,5,6,7,8,8a-octahydro-1,4,-endo, endo-5,8-dimethanonaphthalene)

Data

Pesticide		
WS	0.024 ppm	[6]
Kow	218,000	[6]
BCF—flowing water	4,050	[6]
BCF—static water	1,360	[6]

Discussion

- WS indicates that this chemical could adsorb in soil, be bioaccumulated, and run off with soil particles, and it should not biodegrade and should not be leached.
- Kow indicates that this chemical should bioaccumulate and cause food-chain contamination.
- BCF indicates that this chemical should bioaccumulate and cause food-chain contamination. Residues of the chemical and its by-products could be expected in the food chain.

Exposure

Routes of exposure could be by (1) ingestion of contaminated water and food and (2) inhalation if the chemical were volatile; however, no data are presented on volatility herein.

EPTC (also *S*-Ethyl dipropylthiocarbamate and Eptam)

Data

Pesticide		
WS	365 ppm	[6]
Koc	240	[6]

Discussion

- WS indicates that this chemical could go either way in relation to leaching, runoff, adsorption, biodegradation, and bioaccumulation.
- Koc indicates that the chemical should not be adsorbed to soil.
- Using the values for Koc and WS, it is predicted that this chemical should leach, run off, and be biodegraded, and it should not be adsorbed and should not be bioaccumulated. If it were released into the environment, its residues might be found in the food chain; however, additional data would be needed to confirm this. It is also predicted that these residues should be metabolized in animals.

Exposure

Routes of exposure could be by (1) consumption of contaminated water and food and (2) inhalation if the chemical were volatile; however, no data are presented herein on volatility.

Ethion (*O,O,O′,O′*-tetraethyl *S,S′*-methylene bisphosphorodithioate)

Data

Pesticide		
WS	2 ppm	[6]
Koc	15,400	[6]

Discussion

- WS indicates that this chemical could adsorb to soil, be bioaccumulated, and run off with soil, and it should not leach and should not be biodegraded.

- Koc indicates that this chemical should adsorb to soil and accumulate in soil.

Exposure

Routes of exposure could be by (1) ingestion of contaminated food and water and (2) inhalation if volatility occurred; however, no data are presented herein on volatility.

Ethylbenzene

Data

WS at 20°C	152 mg/L	[4]
Kow	1,413	[4]
VP at 20°C	7 torr	[4]

Discussion

- WS indicates that this chemical could go either way in relation to leaching, runoff, adsorption, biodegradation, and bioaccumulation. Additional data would be needed to make a prediction.
- Kow indicates that bioaccumulation and food-chain contamination could occur. Residues could be expected in the food chain.
- VP indicates volatility; fallout could contaminate food and water. If the chemical were phototransformed, transformation products could fall out and contaminate food and water. Inhalation of the parent chemical and its potential photoproducts could be a problem.

Exposure

Routes of exposure could be by (1) ingestion of contaminated food and water if such occurred, and (2) inhalation of volatile residues.

Ethylene dibromide (EDB)

Data

Pesticide

WS	3,370 ppm	[6]
Koc	44	[6]

Discussion

- WS indicates that leaching, runoff, and biodegradation could occur.
- Koc indicates that food-chain contamination should not occur, nor should adsorption in soil.
- Based on the data, EDB could be mobile, and it could contaminate aquatic environments if not biodegraded.

Exposure

Contamination is possible, of aquatic environments and of air if the chemical is volatile. Exposure could be by (1) ingestion of contaminated drinking water or (2) inhalation if EDB gets into the air.

Ethylene dichloride (also 1,2-Dichloroethane)

Data

WS at 20°C	8,690 mg/L	[4]
Kow	30	[4]
VP at 20°C	61 torr	[4]

Discussion

- WS indicates that this chemical should leach, run off, and be biodegraded, and it should not adsorb to soil and should not be bioaccumulated. The value for Kow supports this prediction.
- Kow indicates that the chemical should not bioaccumulate.
- VP indicates volatility; thus inhalation could be a problem. If it were phototransformed, fallout of photoproducts and the parent chemical could contaminate water and food.
- This chemical could be mobile in the environment, and could contaminate water and food, thus resulting in contamination of the food chain. Inhalation of volatile residues could be a problem.

Exposure

Routes of exposure could be by (1) consumption of contaminated water and food if such occurred, and (2) inhalation of volatile residues.

Fenuron (3-Phenyl-1,1-dimethylurea)

Data

Pesticide		
WS	3,850 ppm	[6]
Koc	27	[6]
Kow	10	[6]

Discussion

- WS indicates that this chemical should leach, run off, and be biodegraded, and it should not be adsorbed in soil and should not be bioaccumulated.
- Koc indicates that this chemical should not adsorb in soil.
- Kow indicates that the chemical should not bioaccumulate.
- This chemical could be mobile in the environment and could cause contamination of the food chain. Residues could dissipate; however, there are no data to discern dissipation.

Exposure

Routes of exposure could be by (1) consumption of contaminated water and food if such occurred, and (2) inhalation if the chemical were volatile; however, no data are presented herein on volatility.

Fluchloralin (*N*-(Chloroethyl)-alpha, alpha, alpha-trifluoro-2,6-dinitro-*N*-propyl-*p*-toluidine) (also Basalin)

Data

Pesticide		
WS	<1 ppm	[6]
Koc	3,600	[6]

Discussion

- WS indicates that this chemical should adsorb to soil, run off with soil, and be bioaccumulated, and it should not be biodegraded and should not be leached.
- Koc indicates that the chemical should adsorb to soil.
- The data indicate that the chemical should adsorb to soil and

bioaccumulate in the food chain if released into the environment. If this occurred, residues in the food chain could be expected.

Exposure

Routes of exposure could be by (1) consumption of contaminated water and food if such occurred, and (2) inhalation if the chemical were volatile; however, no data are presented herein on volatility.

Fluoranthene

Data

WS at 25°C	0.26 mg/L	[4]
Kow	213,796	[4]
VP at 20°C	6×10^{-6} torr	[4]

Discussion

- WS indicates that this chemical could adsorb in soil, run off with soil, and be bioaccumulated, and it should not leach and should not be biodegraded.
- Kow indicates that bioaccumulation and food-chain contamination could occur. Residues could be expected in the food chain.
- VP indicates that volatility should be of no concern.

Exposure

Routes of exposure could be by ingestion of contaminated food and water if such occurred.

Fluorene

Data

WS at 25°C	1.98 mg/L	[4]
Kow	15,136	[4]
VP at 20°C	1.3×10^{-2} torr	[4]

Discussion

- WS indicates that this chemical could adsorb in soil, run off with soil, and be bioaccumulated, and it should not leach and should not be biodegraded.
- Kow indicates that bioaccumulation and food-chain contamination could occur. Residues could be expected in the food chain.
- VP indicates volatility; fallout could contaminate food and water. If the chemical were phototransformed, transformation products also could contaminate food and water. Inhalation of the parent chemical and its potential photoproducts could be a problem.

Exposure

Routes of exposure could be by (1) ingestion of contaminated food and water and (2) inhalation of volatile residues.

Formetanate (*m*-[[(Dimethylamino) methylene] amino] phenyl methylcarbamate) (also Carzol)

Data

Pesticide		
WS	<1,000 ppm	[6]
BCF—static water	0	[6]

Discussion

- WS indicates that this chemical could go either way; however, since it is close to 1,000 ppm and there is a zero BCF, it is predicted that it should leach, run off, and biodegrade, and it should not adsorb in soil and should not be bioaccumulated.
- BCF indicates that there should be no bioaccumulation, and there should be metabolism in animals.
- Although there is no bioaccumulation, and there probably is metabolism in animals, it is predicated that there could be residues in the food chain until some point in time when metabolism is complete.

Exposure

Routes of exposure could be by (1) consumption of contaminated water and food if such occurred, and (2) inhalation if the chemical were volatile; however, no data are presented herein on volatility.

Glyphosate (*N*-(Phosphonomethyl) glycine)

Data

Pesticide		
WS	12,000 ppm	[6]
Koc	2,640	[6]

Discussion

- WS indicates that this chemical could leach, run off and be biodegraded, and it should not bioaccumulate and should not be adsorbed to soil.
- Koc indicates that the chemical could adsorb to soil and accumulate in soil. This contradicts WS; therefore, additional data would be needed for a conclusive prediction.

Exposure

Routes of exposure could by (1) ingestion of contaminated food and water if such occurred, and (2) inhalation if the chemical were volatile; however, no data are presented on volatility herein. Additional data would be needed to clarify the contradictions.

Heptachlor (Heptachlorotetrahydro-4,7-methanoindene and related compounds)

Data

Pesticide		
WS	0.03 ppm	[6]
BCF—flowing water	17,400	[6]
BCF—static water	2,150	[6]

Discussion

- WS indicates that this chemical should adsorb to soil, run off with soil, and be bioaccumulated, and it should not be biodegraded and should not be leached.
- Both BCF values indicate that the chemical should bioaccumulate, and it could contaminate the environment, causing residues in the food chain. The BCF values indicate that little metabolism has occurred.
- This chemical could contaminate the environment if released into it. Food-chain contamination would be expected, with very little metabolism anticipated in the food chain.

Exposure

Routes of exposure could be by (1) consumption of contaminated water and food if such occurred, and (2) inhalation if the chemical were volatile; however, no data are presented herein on volatility.

Hexachlorobenzene

Data

Pesticide		
WS	0.035 ppm	[6]
Koc	3,914	[6]
Kow	168,000	[6]
BCF—flowing water	8,600	[6]
BCF—static water	290	[6]

Discussion

- WS indicates that this chemical could adsorb to soil, be bioaccumulated, and run off with soil particles, and it should not leach and should not be biodegraded.
- Koc indicates that adsorption to soil could occur, and that this chemical could accumulate in soil.
- Kow indicates that this chemical could bioaccumulate and cause food-chain contamination.

- BCF indicates that this chemical could bioaccumulate and cause food-chain contamination. Residues could be expected in the food chain.

Exposure

Routes of exposure could be by (1) ingestion of contaminated food and water if such occurred, and (2) inhalation if the chemical were volatile; however, no data are presented herein on volatility. Residues could be expected in the food chain if environmental contamination occurred.

Hexachlorobutadiene

Data

WS at 20°C	2 mg/L	[4]
Kow	5,495	[4]
VP at 20°C	0.15 torr	[4]

Discussion

- WS indicates that this chemical should be adsorbed in soil, run off with soil, and be bioaccumulated, and it should not leach and should not be biodegraded. The value for Kow supports this prediction.
- Kow indicates that the chemical should bioaccumulate.
- VP indicates that volatility could occur; thus inhalation could be a problem. If the chemical were phototransformed, photoproducts and the parent chemical could fall out and contaminate water and food.
- This chemical could contaminate the food chain because of its volatility. If food-chain contamination occurred, then residues could be expected in water and food which could result in bioaccumulation. Inhalation of volatile residues could be a problem.

Exposure

Routes of exposure could be by (1) consumption of contaminated water and food if such occurred, and (2) inhalation of volatile residues.

Hexachlorocyclopentadiene

Data

WS	0.805 mg/L	[4]
VP at 25°C	0.081 torr	[4]

Discussion

- WS indicates that this chemical could adsorb in soil, run off with soil, and be bioaccumulated, and it should not leach and should not be biodegraded.
- VP indicates that the chemical could be volatile; thus inhalation could be a problem. If it were phototransformed, photoproducts and the parent chemical could fall out and contaminate water and food.
- This chemical could contaminate water and food, resulting in contamination of the food chain. Volatile residues could present an inhalation problem.

Exposure

Routes of exposure could be by (1) consumption of contaminated water and food if such occurred, and (2) inhalation of volatile residues.

Hexachloroethane

Data

WS at 20°C	50 mg/L	[4]
Kow	2,187	[4]
VP at 20°C	0.4 torr	[4]

Discussion

- WS indicates that this chemical could go either way in regard to leaching, runoff, adsorption, biodegradation, and bioaccumulation.
- Kow indicates that the chemical should bioaccumulate. Based on the value for Kow, it is predicted that the chemical could adsorb in soil, run off with soil, and bioaccumulate, and it

should not leach and should not be biodegraded. Additional data would be needed to discern biodegradation and whether metabolism in animals would occur.
- VP indicates that volatility could occur; thus inhalation could be a problem. If the chemical were phototransformed, photoproducts and the parent chemical could fall out and contaminate water and food.

Exposure

Routes of exposure could be by (1) consumption of contaminated water and food if such occurred, and (2) inhalation of volatile residues.

Imidan (also N-(Mercaptomethyl) phthalimide *S*-(*O*,*O*-dimethyl phosphorodithioate), Phosmet, and Prolate)

Data

Pesticide		
WS	25 ppm	[6]
Kow	677	[6]
BCF—static water	11	[6]

Discussion

- WS indicates that the chemical could go either way in regard to leaching, runoff, adsorption, biodegradation, and bioaccumulation.
- Kow also indicates that the chemical could go either way.
- BCF indicates the presence of residues, but not bioaccumulation. This chemical may have been metabolized in animals.
- The data do not agree; however, it is predicted that this chemical could leach, run off, be biodegraded, and be metabolized in animals, and this prediction is based on BCF, and the "in-between" data for WS and Kow. Residues may be found in the food chain, but should dissipate in a short amount of time.

Exposure

Routes of exposure (1) could be by consumption of contaminated water and food if such occurred, and 2) inhalation if the chemical were volatile; however, no data are presented herein on volatility.

Ipazine (also 2-Chloro-4-(diethylamino)-6-(isopropylamine)-*s*-triazine)

Data

Pesticide		
WS	40 ppm	[6]
Koc	1,660	[6]
Kow	2,900	[6]

Discussion

- WS indicates that this chemical could go either way in regard to leaching, runoff, adsorption to soil, biodegradation, and bioaccumulation. Using the WS of 40 ppm, a prediction would be difficult to almost impossible to make without other data.
- Koc indicates that adsorption to soil and accumulation could occur.
- Kow indicates that bioaccumulation, food-chain contamination, and residues in the food chain could occur.

Exposure

Routes of exposure could be by (1) ingestion of contaminated food and water if such occurred, and (2) inhalation if the chemical were volatile; however, no data are presented herein on volatility.

Isocil (5-Bromo-3-isopropyl-6-methyluracil)

Data

Pesticide		
WS	2,150 ppm	[6]
Koc	130	[6]

Discussion

- WS indicates that this chemical could leach, run off, and be biodegraded, and it should not adsorb to soil and should not be bioaccumulated.
- Koc indicates that adsorption to soil should not occur.
- Bioaccumulation should not occur; however, if it reaches the aquatic environment, residues of the chemical may be found in the food chain.

Exposure

Routes of exposure could be by (1) ingestion of contaminated water and food if such occurred, and (2) inhalation if the chemical were volatile; however, no data are presented herein on volatility.

Isopropalin (2,6-Dinitro-*N*,*N*-dipropylcumidine)

Data

Pesticide		
WS	0.11 ppm	[6]
Koc	72,250	[6]

Discussion

- WS indicates that this chemical could adsorb in soil, be bioaccumulated, and run off with soil particles, and it should not leach and should not be biodegraded.
- Koc indicates that this chemical should adsorb to soil and accumulate in it.

Exposure

Routes of exposure could be by (1) ingestion of contaminated food and water if such occurred, and (2) inhalation if the chemical were volatile; however, no data are presented herein on volatility.

Kepone (also Decachlorooctahydro-1,3,4-metheno-2H-cyclobuta [*cd*] pentalen-2-one)

Data

Pesticide		
WS	3 ppm	[6]
BCF	8,400	[6]

Discussion

- WS indicates that kepone should not leach but could run off with soil, and it should not be biodegraded.
- BCF indicates it should bioaccumulate, cause contamination in the food chain, and be persistent in the environment.
- The author ran a computer prediction model at EPA's Athens, Georgia laboratory and predicted that kepone could persist in some aquatic soil environments over 1,000 years.
- Based on the data, bioaccumulation and food-chain contamination could occur, and persistence is long.

Exposure

Consumption of contaminated water and food could be a problem. Inhalation also could be a problem; however, no data are presented herein to indicate volatility.

Lead

Data

Is naturally occurring and has many uses	
Has low water solubility	[4]
Complexes with organic material	[4]
Is not bioaccumulated	[4]
Is methylated by benthic microbes to form tetramethyl lead, which is volatile	[4]

Discussion

- Lead does not break down in the environment; in fact, it can be transformed to higher-molecular-weight compounds (e.g., tetramethyl lead), which are volatile.

- Lead can complex with soils; thus it can be present in clay. Clay tableware, tile roofs, etc., can contain lead. The lead can leach out of clay products, thus contaminating humans and the environment.
- Lead-based paints have contaminated humans.
- Lead solder, used to solder water pipe joints, has leached lead into drinking water.

Exposure

Routes of exposure could be by consumption of products contaminated with lead.

Leptophos (*O*-(4-Bromo-2,5-dichlorophenyl) *O*-methyl phenylphosphonothioate)

Data

Pesticide		
WS	2.4 ppm	[6]
Koc	9,300	[6]
Kow	2,020,000	[6]
BCF—flowing water	750	[6]
BCF—static water	1,440	[6]

Discussion

- WS indicates that this chemical should adsorb to soil, run off with soil, and be bioaccumulated, and it should not leach and should not be biodegraded.
- Koc indicates that adsorption to soil and accumulation could occur.
- Kow indicates that bioaccumulation with potential for food-chain contamination could occur.
- BCF indicates that bioaccumulation with potential for food-chain contamination could occur.
- Residues of this chemical could be expected in the food chain.

Exposure

Routes of exposure could be by (1) ingestion of contaminated food and water if such occurred, and (2) inhalation if the chemical were volatile; however, no data are presented herein on volatility.

Lindane (Gamma isomer of benzene hexachloride)

Data

Pesticide		
WS	0.15 ppm	[6]
Koc	911	[6]
BCF—flowing water	325	[6]
BCF—static water	560	[6]

Discussion

- WS indicates that the chemical could be adsorbed to soil, run off with soil, and be bioaccumulated, and it should not leach and should not be biodegraded.
- Koc indicates that the chemical should not adsorb to soil and should not be accumulated in soil, which contradicts WS.
- BCF indicates potential for bioaccumulation and food-chain contamination. BCF also indicates that residues could be expected in the food chain and that metabolism of the chemical in animals would be very slow.
- Lindane could result in residues in the food chain.

Exposure

Routes of exposure could be by (1) ingestion of contaminated food and water if such occurred, and (2) inhalation if the chemical were volatile; however, no data are presented on volatility herein.

Linuron (3-(3,4-Dichlorophenyl)-1-methoxy-1-methylurea)

Data

Pesticide		
WS	75 ppm	[6]
Koc	820	[6]
Kow	154	[6]
Soil TLC-Rf	0.17	[6]

Discussion

- WS indicates that the chemical could go either way in regard to leaching, runoff, biodegradation, adsorption, and bioaccumulation.
- Koc indicates that adsorption to soil should not occur.
- Kow indicates that bioaccumulation should not occur.
- Soil TLC indicates some adsorption, and that leaching should not occur.
- It is possible that the chemical is biodegraded and has enough adsorption to soil to prevent leaching. It is also possible for a residue to be found in the food chain, based on Kow. Additional data are needed for a more conclusive prediction.

Exposure

Routes of exposure could be by (1) ingestion of contaminated food and water if such occurred, and (2) inhalation if the chemical were volatile; however, no data are presented on volatility herein.

Malathion (O,O-Dimithyl dithiophosphate of diethyl mercaptosuccinate) (also *S*-(1,2-dicarbethoxyethyl) O,O-dimethyl phosophorodithioate)

Data

Pesticide		
WS	145 ppm	[6]
Kow	780	[6]
BCF—static water	0	[6]

Discussion

- WS indicates that the chemical could go either way in regard to leaching, runoff, adsorption to soil, biodegradation, and bioaccumulation.
- Kow indicates that bioaccumulation should not occur.
- BCF indicates that bioaccumulation should not occur.

Exposure

Routes of exposure could be by (1) ingestion of contaminated food and water if such occurred, and (2) inhalation if the chemical were volatile; however, no data are presented herein on volatility.

Methazole (2-(3,4-Dichlorophenyl)-4-methyl-1,2,4-oxadiazolidine-3,5-dione) (also Tunic, VCS-438, Bioxone, Probe, and chlormethazole)

Data

Pesticide		
WS	1.5 ppm	[6]
Koc	2,620	[6]

Discussion

- WS indicates that this chemical should adsorb to soil, run off with soil, and bioaccumulate, and it should not be biodegraded and should not be leached.
- Koc indicates that the chemical could adsorb to soil or not. The value is in the in-between range; however, with the use of the value for WS, it is predicted that the chemical could bioaccumulate, adsorb to soil, and run off with soil.
- The chemical, if released into the environment, could bioaccumulate and cause contamination in the food-chain.

Exposure

Routes of exposure could be by (1) consumption of contaminated water and food if such occurred, and (2) inhalation if the chemical were volatile; however, no data are presented herein on volatility.

Methomyl (S-methyl N-[(methylcarbamoyl) oxy] thioacetimidate)

Data

Pesticide		
WS	10,000 ppm	[6]
Koc	160	[6]
Kow	2	[6]

Discussion

- WS indicates leaching and runoff potential with water and possible biodegradation.
- Koc indicates that soil adsorption should not occur.
- Kow indicates that bioaccumulation and food-chain contamination should not occur, and this is further supported by WS and Koc, and vice versa.
- Based on these data, the chemical could contaminate the aquatic environment if not biodegraded, but it will not bioaccumulate.

Exposure

Exposure could be by consumption of contaminated drinking water or food if such contamination occurred. It should be noted that volatility is of concern, because of direct inhalation. This is not to be interpreted as meaning that the chemical is volatile, as no volatility data are given herein.

Methoxychlor, technical (2,2-Bis (*p*-methoxyphenyl)-1,1,1-trichloroethane 88% and related compounds 12%)

Data

Pesticide		
WS	0.003 ppm	[6]
Koc	80,000	[6]
Kow	47,500	[6]
BCF—flowing water	185	[6]
BCF—static water	1,550	[6]

Discussion

- WS indicates that this chemical could adsorb in soil, run off with soil, and be bioaccumulated, and it should not leach and should not be biodegraded.
- Koc indicates that both adsorption in soil and accumulation in soil could occur.
- Kow indicates that bioaccumulation could occur and cause food-chain contamination.
- BCF indicates that bioaccumulation could occur, which would cause residues in the food chain. If metabolism occurred in animals, it would be slow, according to the BCF values.

Exposure

Routes of exposure could be by (1) ingestion of contaminated food and water if such occurred, and (2) inhalation if the chemical were volatile; however, no data on volatility are presented herein.

2-Methoxy-3,5,6-trichloropyridine

Data

WS	20.9 ppm	[6]
Koc	920	[6]
Kow	18,500	[6]

Discussion

- WS indicates that this chemical could go either way in regard to leaching, runoff, adsorption, biodegradation, and bioaccumulation.
- Koc indicates that the chemical should not adsorb and could be mobile (leach and run off).
- Kow indicates that the chemical should bioaccumulate and could cause food-chain contamination if released into the environment.
- The data are contradictory; however, values for WS and Kow can be used together to make a prediction. It is predicted that this chemical could adsorb to soil, be bioaccumulated, and run

off with soil, and it should not be biodegraded and should not be leached. Kow indicates that food-chain contamination could occur; thus residues could be expected in the food chain.

Exposure

Routes of exposure could be by (1) consumption of contaminated water and food if such occurred, and (2) inhalation if the chemical were volatile; however, no data are presented herein on volatility.

9-Methylanthracene

Data

WS	0.261 ppm	[6]
Koc	65,000	[6]
Kow	117,000	[6]

Discussion

- WS indicates that this chemical should adsorb to soil, run off with soil, and bioaccumulate, and it should not be biodegraded and should not be leached. This prediction is supported by both Koc and Kow values.
- Koc indicates that the chemical should adsorb to soil.
- Kow indicates the chemical should bioaccumulate, and this could cause contamination of the food chain.
- This chemical, if released into the environment, could cause contamination of the food chain.

Exposure

Routes of exposure could be by (1) consumption of contaminated water and food if such occurred, and (2) inhalation if the chemical were volatile; however, no data are presented herein on volatility.

Methyl chloroform (also 1,1,1-Trichloroethane)

Data

WS at 20°C	480 to 4,400 mg/L	[4]
Kow	148	[4]
VP at 20°C	96 torr	[4]

Discussion

- WS has such a large range of values that it would be difficult to predict what could occur in the environment unless other data were available.
- Kow indicates that this chemical should not bioaccumulate. By using the value for Kow and the highest value for WS, the two would be in agreement. The prediction is that the chemical could leach, run off, and biodegrade, and it should not bioaccumulate and should not be adsorbed in soil.
- VP indicates volatility; thus inhalation could be a problem. If the chemical were phototransformed, photoproducts and the parent chemical could fall out and contaminate water and food.
- This chemical should be mobile in the environment; therefore, it could contaminate water and food, and thus could cause residues in the food chain. Volatile residues could be a problem for inhalation. There are no data to discern if dissipation would occur.

Exposure

Routes of exposure could be by (1) consumption of contaminated water and food if such occurred, and (2) inhalation of volatile residues.

3-Methylcholanthrene

Data

WS	0.00323 ± 0.00017 μg/ml	[3]
Kow	2,632,000 ± 701,000	[3]
Koc	1,244,046	[3]

Discussion

- WS indicates that this chemical should not leach or run off unless the runoff is with soil. It indicates bioaccumulation and accumulation.
- Kow indicates bioaccumulation and potential for food-chain contamination.
- Koc indicates adsorption to soil, which should prevent leaching.
- Based on the data, this chemical should not leach, should not be biodegraded, should bioaccumulate, should adsorb to soil, and should be persistent in the environment.

Exposure

This chemical should be kept out of the aquatic environment, as food-chain contamination is probable. If this chemical gets into soil, and crops are planted in the soil, crop uptake of the chemical is likely, as is food-chain contamination. Inhalation could be a problem; however, no data are presented herein on volatility.

Methylene chloride (also Dichloromethane)

Data

WS	13,2000 to 20,000 mg/L	[4]
Kow	17.78	[4]
VP at 20°C	362.4 torr	[4]

Discussion

- WS indicates that this chemical should leach, run off, and be biodegraded, and it should not adsorb to soil and should not be bioaccumulated.
- Kow indicates that the chemical should not bioaccumulate.
- VP indicates that this chemical is volatile, and this could cause inhalation problems. If the chemical were phototransformed, photoproducts and the parent chemical could fall out onto water and food, and thus could contaminate the food chain.

- This chemical should leach and run off into aquatic environments; however, this may not be a problem, because the chemical is very volatile. It should not last very long in any environment, because of its volatility. Biodegradation also should prevent this chemical from lasting long in the environment. The major concern should be its inhalation and phototransformation, if that occurred, and fallout of photoproducts onto water and soil. No data are presented herein on photo aspects.

Exposure

Route of exposure should be by inhalation of volatile residues.

Methyl isothiocyanate

Data

Pesticide

WS	7,600 ppm	[6]
Koc	6	[6]

Discussion

- WS indicates that this chemical could leach, run off, and be biodegraded, and it should not adsorb in soil and should not be bioaccumulated.
- Koc indicates that adsorption in soil should not occur; therefore, accumulation in soil also should not occur.

Exposure

Routes of exposure could be by (1) ingestion of contaminated food and water if such occurred, and (2) inhalation if the chemical were volatile; however, no data are presented herein on volatility.

2-Methylnaphthalene

Data

WS	25.4 ppm	[6]
Koc	8,500	[6]
Kow	13,000	[6]

Discussion

- WS indicates that this chemical could go either way in regard to leaching, runoff, adsorption, biodegradation, and bioaccumulation.
- Koc indicates the chemical could adsorb to soil, and this is supported by the Kow value.
- Kow indicates that this chemical should bioaccumulate, and this could cause food-chain contamination.
- If this chemical were released into the environment, it could contaminate the food chain.

Exposure

Routes of exposure could be by (1) consumption of contaminated water and food if such occurred, and (2) inhalation if the chemical were volatile; however, no data are presented herein on volatility.

Methylparathion (*O,O*-Dimethyl-*O*-*p*-nitrophenyl phosphorothioate)

Data

Pesticide		
WS	57 ppm	[6]
Koc	9,800	[6]
Kow	82	[6]
BCF—static water	95	[6]
VP	9.5 mm Hg $\times$ 10^6 at 20°C	[3]
Degrades in soil		[3]

Discussion

- WS of this chemical could pose a problem of mobility, but this is unlikely because of its Koc.
- Koc indicates that adsorption is likely.
- Kow indicates residues; however, bioaccumulation is unlikely.
- BCF indicates a residue; however, bioaccumulation is unlikely.
- VP indicates volatility, and this could be an inhalation problem.

- Biodegradation by soil microbes is assumed, as this chemical degrades in soil; however, no mention is made herein of whether soil degradation was by hydrolysis or other means.
- Based on the data, this chemical could get into air because of its VP. Bioaccumulation in the food-chain and accumulation in the soil are unlikely because this chemical is degraded in soil. Kow indicates it would be metabolized in animals.

Exposure

Two possible routes of exposure are (1) inhalation if one is in a contaminated area, and (2) ingestion only if this chemical gets into drinking water or on food that is consumed. Volatility is of concern, and fallout to noncontaminated areas is possible. No data are presented herein on phototransformation, and this possibility should be considered.

Metobromuron (3- (*p*-Bromophenyl) -1-methoxy-1-methylurea) (also Patoran)

Data

Pesticide		
WS	330 ppm	[6]
Koc	60	[6]

Discussion

- WS indicates that this chemical could go either way in regard to leaching, runoff, adsorption, biodegradation, and bioaccumulation.
- Koc indicates that the chemical should not adsorb.
- Based on the value for Koc and the lower end of the middle range for WS, it is predicted that this chemical could be mobile, and this could result in residues in the food chain.

Exposure

Routes of exposure could be by (1) consumption of contaminated water and food if such occurred, and (2) inhalation if the chemical were volatile; however, no data are presented herein on volatility.

Metolachlor (2-Chloro-N- (ethyl-6-methylphenyl) -N- (2-methoxy-1-methlethyl) acetamide)

Data

Pesticide		
Stable to hydrolysis	$t_{1/2}$ 200 days at 30°C	[9]
Stable to photolysis		[9]
Biodegraded by soil microbes		[9]
Has been shown to break down to over 20 chemicals		[9]
BCF	<10	[9]

Discussion

- Biodegradation by soil microbes is indicated; therefore, the chemical should be water-soluble, could leach, and could run off.
- BFC indicates that bioaccumulation is unlikely.
- The chemical is stable to phototransformation; thus it would be stable in air, if volatile, unless it were biodegraded.

Exposure

If microbial populations do not rapidly biodegrade the chemical, then contamination of the aquatic environment is likely. Consumption of contaminated drinking water could be the route of exposure. Contaminated water could contaminate the food chain. Inhalation could be a problem; however, no data are presented herein on volatility.

Mexacarbate (4-Dimethylamino)-3, 5-xylyl methylcarbamate) (also Zectran)

Data

Pesticide		
WS	120 ppm	[6]
Kow	1,370	[6]

Discussion

- WS indicates that the chemical could go either way in regard to leaching, runoff, adsorption, biodegradation, and bioaccumulation.
- Kow indicates that the chemical should bioaccumulate.
- If this chemical were released into the environment, it could bioaccumulate and could cause contamination in the food chain. There are no data to indicate metabolism in animals; therefore, residues could be expected in the food chain.

Exposure

Routes of exposure could be by (1) consumption of contaminated water and food if such occurred, and (2) inhalation if the chemical were volatile; however, no data are presented herein on volatility.

Mirex (also Dodecachlorooctahydro-1,3,4-methano-1*H*-cyclobuta [*cd*] pentalene)

Data

Pesticide		
WS	0.6 ppm	[6]
BCF—static water	220	[6]

Discussion

- WS indicates that this chemical could adsorb in soil, run off with soil, and be bioaccumulated, and it should not leach and should not be biodegraded.

- BCF indicates potential for bioaccumulation and food-chain contamination. Residues could be expected in the food chain.

Exposure

Routes of exposure could be by (1) ingestion of contaminated water and food if such occurred, and (2) inhalation if the chemical were volatile; however, no data are presented herein on volatility.

Monolinuron (also 3-(*p*-Chlorophenyl)-1-methoxy-1-methylurea)

Data

Pesticide		
WS	580 ppm	[6]
Koc	200	[6]
Kow	40	[6]

Discussion

- WS indicates that this chemical could go either way in regard to leaching, runoff, adsorption, biodegradation, and bioaccumulation.
- Koc indicates that the chemical should not adsorb to soil.
- Kow indicates that the chemical should not bioaccumulate.
- Using the three above values, it can be predicted that this chemical should leach, run off, and be biodegraded, and it should not be bioaccumulated and should not adsorb to soil. It is predicted that this chemical is mobile; thus food-chain contamination could occur. The $t_{1/2}$ for biodegradation and metabolism cannot be predicted; therefore, one may predict that residues could be in the food chain, but additional data would be needed for a more conclusive prediction.

Exposure

Routes of exposure could be by (1) consumption of contaminated water and food if such occurred, and (2) inhalation if the chemical were volatile; however, no data are presented herein on volatility. Additional data would be needed on biodegradation and metabolism to predict whether residues would be a problem in the food chain.

Monuron (3-(*p*-Chlorophenyl-1-dimethylurea) (also Telvar)

Data

Pesticide		
WS	230 ppm	[6]
Koc	100	[6]
Kow	29	[6]

Discussion

- WS indicates that this chemical could go either way in regard to leaching, runoff, adsorption, biodegradation, and bioaccumulation.
- Koc indicates that the chemical should not adsorb.
- Kow indicates that the chemical should not bioaccumulate.
- The data are contradictory, and no two pieces agree. No prediction can be made.

Exposure

Routes of exposure cannot be predicted without additional data. However, to be on the safe side, if a prediction must be given, then it could be that consumption of contaminated water and food, as well as inhalation, could be a problem.

Naphthalene

Data

WS	31.7 ppm	[6]
Koc	1,300	[6]
Kow	2,040	[6]

Discussion

- WS indicates that this chemical is in the range of nonmobile to mobile.

- Koc indicates that adsorption to soil should occur, and solubility should not be conducive to leaching.
- Kow indicates bioaccumulation and food-chain contamination potential, and a water solubility that should not be conducive to mobility (leaching).
- Based on the data, the chemical should not leach, but it could run off with soil particles to contaminate aquatic environments. If any water is contaminated, or if food crops are grown in soils containing this chemical, then bioaccumulation and food-chain contamination are likely.

Exposure

Exposure would be by consumption of contaminated water or food if such occurred. Volatility data are not given; but this chemical could be a problem if volatile, if it were inhaled.

1-Naphthol

Data

WS	866 ± 31 μg/ml	[3]
Kow	700 ± 62	[3]
Koc	522	[3]
Soil TLC-Rf	0	[3]

Discussion

- WS indicates that leaching should not occur, but adsorption to soil should occur, runoff with soil particles could occur, and bioaccumulation is possible.
- Kow indicates that bioaccumulation could occur.
- Koc indicates adsorption to soil.
- Soil TLC-Rf indicated no leaching and showed potential for soil adsorption, bioaccumulation, and low solubility.
- Based on the data, the chemical should not leach, should not biodegrade, and should bioaccumulate.

Exposure

Exposure could be by consumption of contaminated water or food if such contamination occurred. Another route could be by inhalation if the chemical were volatile, but no data are given herein on volatility.

Nitralin (also 4-(Methylsulfonyl)-2,6-dinitro-*N*,*N*-dipropylaniline and Planavin)

Data

Pesticide		
WS	0.6 ppm	[6]
Koc	960	[6]

Discussion

- WS indicates that this chemical should adsorb to soil, run off with soil, and be bioaccumulated, and it should not be biodegraded and should not be leached.
- Koc indicates that the chemical should not be adsorbed, and this is not in agreement with the value for WS. The Koc value is very close to the value for going either way; perhaps the chemical could do that.
- In this situation, where the two pieces of data are not in agreement, and it is predicted that the Koc value could go either way, it should be predicted that the chemical could adsorb, and that, if released into the environment, it could bioaccumulate and cause food-chain contamination. Additional data would be needed to support this; but when one must make a prediction on limited data, sometimes it can be done.

Exposure

Routes of exposure could be by (1) consumption of contaminated water and food if such occurred, and (2) inhalation if the chemical were volatile; however, no data are presented herein on volatility. Additional data would be needed for a more conclusive prediction on food-chain contamination.

Nitrapyrin (2-Chloro-6-(trichloromethyl) pyridine) (also N-Serve)

Data

Pesticide		
WS	40 ppm	[6]
Koc	420	[6]
Kow	2,590	[6]

Discussion

- WS indicates that this chemical could go either way in regard to leaching, runoff, adsorption, biodegradation, and bioaccumulation.
- Koc indicates that this chemical should not be adsorbed and could be mobile.
- Kow indicates that the chemical should bioaccumulate and could cause food-chain contamination.
- WS and Kow are in close agreement; however, Koc does not agree. It is predicted that the chemical could bioaccumulate and cause food-chain contamination. Residues could be expected in the food chain. With the predictions on food-chain contamination, mobility can also be predicted. Biodegradation and metabolism cannot be predicted without additional data.

Exposure

Routes of exposure could be by (1) consumption of contaminated water and food if such occurred, and 2) inhalation if the chemical were volatile; however, no data are presented herein on volatility.

Nitrobenzene

Data

Pesticide		
WS	1,780 ppm	[6]
Kow	62	[6]
BCF—static water	29	[6]

Discussion

- WS indicates that this chemical could leach, run off, and be biodegraded, and it should not adsorb in soil and should not be bioaccumulated.
- Kow indicates that bioaccumulation should not occur.
- BCF indicates that bioaccumulation should not occur.
- BCF indicates that residues could be expected in the food chain.

Exposure

Routes of exposure could be by (1) ingestion of contaminated food and water if such occurred, and 2) inhalation if the chemical were volatile; however, no data are presented herein on volatility.

2-Nitrophenol

Data

WS at 20°C	2,100 mg/L	[4]
Kow	58	[4]
VP at 49.3°C	1 torr	[4]

Discussion

- WS indicates that this chemical could leach, run off, and be biodegraded, and it should not adsorb in soil and should not be bioaccumulated.
- Kow indicates that bioaccumulation should not be a problem.
- VP indicates volatility; fallout could contaminate food and water. If the chemical were phototransformed, transformation products could contaminate food and water. Inhalation of the parent chemical and its potential photoproducts could be a problem.

Exposure

Routes of exposure could be by (1) ingestion of contaminated food and water and (2) inhalation of volatile residues.

4-Nitrophenol

Data

WS at 25°C	16,000 mg/L	[4]
Kow	81	[4]
VP at 146°C	2.2 torr	[4]

Discussion

- WS indicates that this chemical could leach, run off, and be biodegraded, and it should not adsorb in soil and should not be bioaccumulated.
- Kow indicates that bioaccumulation should not occur.
- VP indicates volatility; fallout could contaminate food and water. If the chemical were phototransformed, transformation products could fall out and contaminate food and water. Inhalation of the parent chemical and its potential photoproducts could be a problem.

Exposure

Routes of exposure could be by (1) ingestion of contaminated food and water and (2) inhalation of volatile residues.

Norflurazon (4-Chloro-5-(methylamino)-2-(alpha, alpha, alpha-trifluroro-*m*-tolyl)-3(2H)-pyridazinone) (also San 9789, Zorial, Monometflurazon, and Evital)

Data

Pesticide

WS	28 ppm	[6]
Koc	1,914	[6]
Soil TLC-Rf	0.4	[1]

Discussion

- WS indicates that this chemical could go either way in regard to leaching, runoff, adsorption, biodegradation, and bioaccumulation.

- Koc indicates that the chemical should adsorb to soil and run off with soil.
- Soil TLC indicates that leaching of this chemical should not occur.
- By using Koc and soil TLC values, the value for WS can then be used to predict that this chemical should adsorb to soil, run off with soil, and bioaccumulate, and it should not biodegrade and should not leach. All three values then will support the prediction that bioaccumulation could occur and cause food-chain contamination. Residues could be expected in the food chain if this chemical were released into the environment.

Exposure

Routes of exposure could be by (1) consumption of contaminated water and food if such occurred, and (2) inhalation if the chemical were volatile; however, no data are presented herein on volatility.

Oxadiazon (2-*tert*-Butyl-4-(2,4-dichloro-5-isopropoxyphenyl)-delta^2-1,3,4-oxadiazolin-5-one) (also Ronstar)

Data

Pesticide

WS	0.7 ppm	[6]
Koc	3,241	[6]

Discussion

- WS indicates that this chemical should adsorb to soil, run off with soil, and be bioaccumulated, and it should not leach and should not be biodegraded.
- Koc indicates that the chemical should adsorb to soil, and this supported is by the value for WS.
- This chemical could bioaccumulate and cause food-chain contamination if released into the environment. If food-chain contamination occurred, then residues could be expected in the food chain.

Exposure

Routes of exposure could be by (1) consumption of contaminated water and food if such occurred, and (2) inhalation if the chemical were volatile; however, no data are presented herein on volatility.

Paraquat

Data

Pesticide		
WS	1,000,000 ppm	[6]
Koc	15,473	[6]
Soil TLC-Rf	0 to 0.13	[5]

Discussion

- WS indicates that this chemical could leach, run off, and be biodegraded, and it should not adsorb in soil and should not be bioaccumulated. Koc and soil TLC contradict WS predictions.
- Koc indicates that adsorption in soil should occur for this chemical; however, WS does not indicate this.
- Soil TLC indicates that this chemical should not leach and should be adsorbed in soil.
- This chemical does not adsorb to the organic matter of soils, but it is known to exchange with the soils' cation exchange capacity. Something had to cause the difference found with WS, Koc, and soil TLC, but only an experienced person or a person with extra knowledge of this chemical would know about the cation exchange. However, one could predict that adsorption was not based on adsorption to organic matter or organic carbon, as discussed above in the section on "Soil Sorption" in Chapter 2. Adsorption of some form did occur, however, as indicated by Koc and soil TLC.
- Bioaccumulation could not be predicted based on Koc. One can predict that accumulation in soil could occur, and that there is potential for residues in food and water using WS, Koc, and soil TLC.

Exposure

Routes of exposure could be by (1) ingestion of contaminated food and water if such occurred, and (2) inhalation if the chemical were volatile; however, no data are presented herein on volatility. Additional data would be needed to confirm these predictions.

Parathion (O,O-Diethyl O-*p*-nitrophenyl phosphorothioate)

Data

Pesticide

WS	24 ppm	[6]
Koc	4,800	[6]
Kow	6,400	[6]
BCF—static water	335	[6]

Discussion

- WS indicates that this chemical could adsorb in soil, run off with soil, and be bioaccumulated, and it should not leach and should not be biodegraded.
- Koc indicates that adsorption in soil could occur, and that there is potential for accumulation in soil.
- Kow indicates that bioaccumulation could occur.
- BCF indicates that bioaccumulation and food-chain contamination could occur. Residues could be expected in the food chain.

Exposure

Routes of exposure could be by (1) ingestion of contaminated food and water if such occurred, and (2) inhalation if the chemical were volatile; however, no data are presented herein on volatility.

Pentachlorobenzene

Data

WS	0.135 ppm	[6]
Kow	154,000	[6]
BCF—flowing water	about 5,000	[6]

Discussion

- WS indicates that this chemical could adsorb in soil, run off with soil, and be bioaccumulated, and it should not leach and should not be biodegraded.
- Kow indicates that bioaccumulation could occur.
- BCF indicates that bioaccumulation and food-chain contamination could occur. Residues could be expected in the food chain.

Exposure

Routes of exposure could be by (1) ingestion of contaminated food or water if such occurred, and (2) inhalation if the chemical were volatile; however, no data are presented herein on volatility.

Pentachlorophenol

Data

Pesticide		
WS	14 ppm	[6]
Koc	900	[6]
Kow	102,000	[6]
VP	0.00011 torr (calculated)	[4]
Adsorbs light	245 nm and 318 nm	[4]

Discussion

- WS is in the "either way" range.
- Koc indicates that sorption can occur.
- Kow indicates that bioaccumulation should occur; thus food-chain contamination is likely.
- VP indicates that volatility is not a problem.
- This chemical adsorbs light; therefore, phototransformation can occur in water and on surfaces.

Exposure

Exposure could be by consumption of contaminated water or food if such occurred. Inhalation could be a problem if the chemical were volatile; however, volatility is unlikely.

Phenanthrene

Data

WS	1.29 ppm	[6]
Koc	23,000	[6]
Kow	32,900	[6]
Kow	28,840	[4]
VP at 20°C	6.8×10^{-4} mm Hg	[4]

Discussion

- WS indicates that this chemical should be adsorbed to soil, run off with soil, and be bioaccumulated, and it should not be biodegraded and should not be leached.
- Koc indicates that the chemical should be adsorbed to soil and could run off with soil.
- Kow indicates that the chemical should bioaccumulate and could cause food-chain contamination.
- VP indicates potential for volatility and thus inhalation problems. If the chemical were volatile, phototransformation products, if formed, along with the parent chemical, could fall out and contaminate water and the food chain.
- The four pieces of data complement each other. This chemical, if released in the environment, could bioaccumulate and cause food-chain contamination. Residues could be expected in the food chain. Volatile residues of the parent chemical and its possible photoproducts also could present an inhalation problem and a contamination problem, were they to fall out onto water and the food chain.

Exposure

Routes of exposure could be by (1) consumption of contaminated water and food if such occurred, and (2) inhalation; however, additional data are needed for a complete exposure picture.

Phenol (also Carbolic acid)

Data

WS	82,000 ppm	[6]
WS	93,000 ppm	[4]

Koc	27	[6]
VP	0.02 mm Hg	[7]
VP (supercooled liquid)	0.5293 torr	[4]
Kow	1.46	[4]

Discussion

- The WS of 82,000 to 93,000 ppm is an indication that phenol is water-soluble, and should leach into groundwater, run off into surface water, and be biodegraded.
- Koc indicates that adsorption in soil is likely.
- Based on its volatility, the chemical should volatilize from H_2O and soil surfaces.
- Kow indicates that bioaccumulation is unlikely.
- Based on its solubility, the chemical should not adsorb to soil, should be degraded by microbes, and should not bioaccumulate. This is supported by a low Koc of 27 and a low Kow of 1.46. Volatility could be a problem if the chemical were inhaled.

Exposure

Two routes of exposure are possible: (1) inhalation, because of the chemical's volatility, and (2) consumption of contaminated drinking water.

Phosalone (S-[6-Chloro-3-mercaptomethyl)-2-benzoxazolinone] O,O-diethyl phosphorodithioate) (also Zolone)

Data

Pesticide

WS	10 ppm	[6]
Kow	20,100	[6]

Discussion

- WS indicates that this chemical should be adsorbed to soil, be bioaccumulated, and run off with soil, and it should not be biodegraded and should not be leached.
- Kow indicates that the chemical should be bioaccumulated and could cause food-chain contamination.

- The data support a prediction that the chemical should bioaccumulate and could contaminate the food chain. If this occurred, residues could be expected in the food chain.

Exposure

Routes of exposure could be by (1) consumption of contaminated water and food, and (2) inhalation if the chemical were volatile; however, no data are presented herein on volatility.

Phthalic anhydride

Data

WS	6,200 ppm	[6]
Kow	0.24	[6]
BCF—static water	0	[6]

Discussion

- WS indicates that this chemical should leach, run off, and be biodegraded, and it should not adsorb to soil and should not be bioaccumulated.
- Kow indicates that this chemical should not be bioaccumulated and should not cause food-chain contamination.
- BCF indicates that the chemical should not bioaccumulate.
- The three pieces of data support each other. This chemical should not bioaccumulate and should not contaminate the food chain. Residues should not be expected in the food chain. If the chemical were released in water, then consumption of the water could be a problem.

Exposure

Routes of exposure could be by (1) consumption of contaminated water if such occurred, and (2) inhalation if the chemical were volatile; however, no data are presented herein on volatility.

Phorate (O,O-Diethyl S-[(ethylthio) methyl] phosphorodithioate)

Data

Pesticide		
WS	50 ppm	[6]
Koc	3,200	[6]

Discussion

- WS indicates that the chemical could go either way in regard to leaching, runoff, biodegradation, adsorption, and bioaccumulation.
- Koc indicates that adsorption to soil could occur. With this piece of data one could predict potential for bioaccumulation and food-chain contamination. Residues could be expected in the food chain.

Exposure

Routes of exposure could be by (1) ingestion of contaminated food and water if such occurred, and (2) inhalation if the chemical were volatile; however, no data are presented herein on volatility.

Picloram (4-Amino-3,5,6-trichloropicolinic acid) (also Tordon)

Data

Pesticide		
WS	430 ppm	[6]
Koc	17	[6]
Kow	2	[6]
BCF—static water	0.02	[6]
Soil TLC-Rf	0.84	[1]

Discussion

- WS, supported by Koc and Kow, indicates that this chemical could leach, run off, and be biodegraded, and it should not adsorb in soil and should not be bioaccumulated.

- Koc indicates that adsorption to soil should not be a problem.
- Kow indicates that bioaccumulation should not be a problem.
- BCF indicates that bioaccumulation should not be a problem.
- Soil TLC indicates that leaching is possible.
- This chemical could leach, run off, and contaminate the food chain if biodegradation were not rapid. Since no data are presented on biodegradation in soil, it is predicted that leaching and runoff could occur and result in food-chain contamination.

Exposure

Routes of exposure could be by (1) ingestion of contaminated food and water if such occurred, and (2) inhalation if the chemical were volatile; however, no data are presented herein on volatility.

Polychlorinated biphenyls (also PCB's and Aroclors)

Data

WS	0.0027 to 0.34 ppm	[4]
VP	0.0000771 to 0.00406 mm Hg	[4]
Kow	647 to 21,700	[4]
Absorbs light	280 to 320 nm range	[4]
Does not hydrolyze		[4]

Discussion

- Polychlorinated biphenyls, as a group, are benzenoid hydrocarbons with different numbers of chlorine atoms on different parts of the molecule. Aroclors, or PCB's, are mixtures of biphenyl molecules with different numbers of chlorine atoms attached to the molecule.
- WS indicates that they should not leach, should be adsorbed to soil organic matter, should not be biodegraded, and could bioaccumulate.
- Kow indicates that bioaccumulation and food-chain contamination are likely.
- VP indicates that volatility should not be a problem, and that phototransformation is likely.

- Absorption of light indicates that phototransformation is likely.
- Based on the data, these chemicals should be persistent in the environment, accumulate in soils, and bioaccumulate. Food-chain contamination is likely.

Exposure

The main route of exposure should be through consumption of contaminated drinking water and food if that occurred.

Profluralin (*N*-(Cyclopropylmethly)-alpha, alpha, alpha-trifluoro-2,6-dinitro-*N*-propyl-*p*-toluidine) (also Preqard and Tolban)

Data

Pesticide		
WS	0.1 ppm	[6]
Koc	8,600	[6]

Discussion

- WS indicates that this chemical should adsorb in soil, run off with soil, and be bioaccumulated, and it should not leach and should not be biodegraded.
- Koc indicates that the chemical could go either way; however, by using the value given for WS, it is predicted that the chemical could adsorb in soil.
- If this chemical were released into the environment, it could bioaccumulate and cause contamination of the food chain.

Exposure

Routes of exposure could be by (1) consumption of contaminated water and food if such occurred, and (2) inhalation if the chemical were volatile; however, no data are presented herein on volatility.

Propham (also Isopropyl carbanilate, IPC, and prophos)

Data

Pesticide		
WS	250 ppm	[6]
Koc	51	[6]
Soil TLC-Rf	0.51	[1]

Discussion

- WS indicates that this chemical could go either way in regard to leaching, runoff, adsorption, biodegradation, and bioaccumulation.
- Koc indicates that the chemical should not be adsorbed to soil and could be mobile.
- Soil TLC indicates that the chemical could leach.
- Using all three pieces of data, one can predict that this chemical could leach, run off, and be biodegraded, and it should not be adsorbed and should not be bioaccumulated. If it were released into the environment, residues could be found in water and in the food chain, but not be bioaccumulated in animals. No data are presented on $t_{1/2}$ in soils or in animals; thus one cannot predict whether residues could be present in the food chain.

Exposure

Routes of exposure could be by (1) consumption of contaminated water and food if such occurred, and (2) inhalation if the chemical were volatile; however, no data are presented herein on volatility.

Propoxur (also *o*-Isopropoxyphenyl methylcarbamate, Baygon, Sendran, Tendex Suncide, Bay 9010, etc.)

Data

Pesticide		
WS	2,000 ppm	[6]
Koc	33	[6]
BCF—static water	146	[6]

Discussion

- WS indicates that this chemical should leach, run off, and be biodegraded, and it should not be adsorbed and should not be bioaccumulated.
- Kow indicates that the chemical should not bioaccumulate, and this in agreement with values given for WS and BCF.
- BCF indicates that the chemical should not bioaccumulate; however, contamination of the food chain and residues in the food chain could be expected.
- The data are in agreement. This chemical should not bioaccumulate. Contamination of the food chain could be expected if the chemical were released into the environment. If this occurred, residues could be expected in the food chain; however, metabolism should reduce the amount of residue present.

Exposure

Routes of exposure could be by (1) consumption of contaminated water and food if such occurred, and (2) inhalation if the chemical were volatile; however, no data are presented herein on volatility.

Pyrazon (5-Amino-4-chloro-2-phenyl-3(2*H*)-pyridazinone) (also Pyramin (*e*))

Data

Pesticide		
WS	400 ppm	[6]
Koc	120	[6]
Soil TLC-Rf	0.44	[1]

Discussion

- WS indicates that this chemical could go either way in regard to leaching, runoff, adsorption, biodegradation, and bioaccumulation.
- Koc indicates that the chemical should not be adsorbed and could be mobile.

- Soil TLC indicates that the chemical could go either way with respect to adsorption and leaching.
- The data for this chemical are inclusive; however, the prediction is that the chemical could leach, run off, and be biodegraded. Some adsorption to soil could occur. Contamination of the food chain is possible; however, bioaccumulation should not occur. If the chemical were released into the environment, residues could be expected in the food chain, but not bioaccumulation.

Exposure

Routes of exposure could be by (1) consumption of contaminated water and food if such occurred, and (2) inhalation if the chemical were volatile; however, no data are presented herein on volatility.

Pyrene

Data

WS	$0.135 \pm 0.013\ \mu g/ml$	[4]
Kow	$124{,}000 \pm 11{,}000$	[4]
Koc	63,400	[4]
VP	6.85×10^{-7} torr	[4]
Photolysis absorbs solar radiation		[4]

Discussion

- WS indicates that leaching and biodegradation are unlikely; however, soil adsorption is likely.
- Koc indicates that runoff is unlikely unless with soil particles, and Kow further supports the prediction that leaching is unlikely. Koc indicates soil adsorption.
- Kow indicates that bioaccumulation and food-chain contamination may occur, and that prediction is supported by WS and Koc.
- Volatility should not be a problem.

- The chemical should not be mobile, could bioaccumulate, and could cause food-chain contamination.

Exposure

Routes of exposure could be by (1) drinking contaminated water and (2) eating contaminated plants and animals. These routes of exposure could occur if the chemical were spilled on cropland or in the aquatic environment.

Pyroxychlor (2-Chloro-6-methoxy-4-(trichloromethyl) pyridine)

Data

Pesticide		
WS	11.3 ppm	[6]
Koc	3,000	[6]
BCF—static water	239	[6]

Discussion

- WS, supported by Koc and BCF, indicates that this chemical could adsorb to soil, run off with soil, and be bioaccumulated, and it should not leach and should not be biodegraded. This could not have been predicted on WS alone because it could gone either way. However, with the use of Koc and BCF, such prediction is possible.
- Koc indicates that adsorption to soil could occur.
- BCF indicates that residues could be expedited in the food chain, as well as bioaccumulation potential.

Exposure

Routes of exposure could be by (1) ingestion of contaminated food and water if such occurred, and (2) inhalation if the chemical were volatile; however, no data are presented herein to indicate volatility.

Radon

Data

Soluble in water	[8]
Gas	[8]

Discussion

- WS indicates mobility in soil and the potential for aquatic contamination.
- Radon is a gas and can move through soil to contaminate the atmosphere.

Exposure

Two routes of exposure could be: (1) inhalation and (2) consumption of contaminated drinking water.

Ronnel (*O*,*O*-Dimethyl *O*-(2,4,5-trichlorophenyl) phosphorothioiate) (also Trolene, Korlan, fenchlorphos, and Viozene)

Data

Pesticide

WS	6 ppm	[6]
Kow	46,400	[6]
VP at 25°C	8×10^{-4} mm Hg	[8]

Discussion

- WS indicates that this chemical should not leach and should not be biodegraded, and it should be adsorbed to soil, be bioaccumulated, and run off with soil.
- Kow indicates that the chemical should bioaccumulate, and this is in agreement with the value for WS.
- VP indicates that the chemical could be volatile; thus inhalation could be a problem. No data are presented to indicate phototransformation. Fallout onto water and food could be a problem.

- This chemical should bioaccumulate, and residues could be expected in the food chain. The chemical could be volatile enough to fall out onto water and food, causing food-chain contamination. Volatility also presents an inhalation problem. No data are presented to indicate whether phototransformation could occur. No data are presented on metabolism in animals so that one could discern $t_{1/2}$ in animals.

Exposure

Routes of exposure could be by (1) consumption of contaminated water and food and (2) inhalation of volatile residues.

Silvex (2-(2,4,5-Trichlorophenoxy) propionic acid)

Data

Pesticide		
WS	140 ppm	[6]
Kow	2,600	[6]

Discussion

- WS, supported by Kow, indicates that this chemical could adsorb in soil, run off with soil, and be bioaccumulated, and it should not leach and should not be biodegraded. WS alone could not be used to predict this, as any of the above could go either way, based on WS. Additional data should be used for this prediction.
- Kow indicates that bioaccumulation and food-chain contamination could occur. Residues could be expected in the food chain.

Exposure

Routes of exposure could be by (1) ingestion of contaminated food and water if such occurred, and (2) inhalation if the chemical were volatile; however, no data are presented herein to indicate volatility.

Simazine (2-Chloro-4,6-bis (ethylamino)-*s*-triazine)

Data

Pesticide		
WS	3.5 ppm	[6]
Koc	135	[6]
Kow	155	[6]
BCF—flowing water	1	[6]
Soil TLC-Rf	0.45	[1]
Soil TLC-Rf	0.31 to 0.96	[5]

Discussion

- WS indicates that this chemical could leach, run off, and be biodegraded, and it should not adsorb in soil and should not be bioaccumulated.
- Koc indicates that adsorption to soil should be of no concern.
- Kow indicates that bioaccumulation should not occur; however, residues could be expected in the food chain.
- BCF indicates that bioaccumulation should not occur; however, residues could be found in the food chain.
- Soil TLC indicates that leaching could occur in some soils.

Exposure

Routes of exposure could be by (1) ingestion of contaminated food and water if such occurred, and (2) inhalation if the chemical were volatile; however, no data are presented herein to indicate volatility.

2,4,5-T (also 2,4,5-Trichlorophenoxyacetic acid)

Data

Pesticide		
WS	238 ppm	[6]
Koc	53	[6]
Kow	4	[6]
BCF—static water	25	[6]
Soil TLC-Rf	0.54	[1]

Discussion

- WS, supported with Koc and Kow, indicates that this chemical could leach, run off, and be biodegraded, and it should not adsorb to soil and should not be bioaccumulated.
- Koc indicates that adsorption in soil should be of no concern.
- BCF indicates that bioaccumulation should be of no concern; however, residues could be expected in the food chain.
- Soil TLC indicates the potential to leach.

Exposure

Routes of exposure could be by (1) ingestion of contaminated food and water if such occurred, and (2) inhalation if the chemical were volatile; however, no data are presented herein to indicate volatility.

TCE (also Trichloroethylene and Ethylene trichloride)

Data

WS at 20°C	1,100 ppm	[8]
Kow	195	[4]
VP at 20°C	57.9 torr	[4]
Does not absorb visible or near UV		[4]
Photooxidation occurs		[4]

Discussion

- WS indicates that this chemical could leach, run off, and be biodegraded, and it should not adsorb to soil and should not be bioaccumulated.
- Kow indicates that this chemical should not bioaccumulate.
- VP indicates this chemical could be volatile, and it should not be phototransformed because it does not absorb visible or UV light. Fallout onto aquatic and food environments is possible; thus contamination of the food chain could occur.
- Photooxidation has occurred, and TCE has photooxidized to form dichloroacetyl chloride and phosgene. *Note:* These

breakdown products are mentioned because phosgene is a highly toxic gas that will produce death.
- Contamination of the food chain is possible, and residues in the food chain are possible.

Exposure

Routes of exposure could be by (1) ingestion of contaminated food and water if such occurred, and (2) inhalation of volatile residues.

Terbicil (3-*tert*-Butyl-5-chloro-6-methyl uracil)

Data

Pesticide		
WS	0.071 g/100 ml at 25°C	[2]
Soluble in water		[2]
Mobile in soil because of low adsorption		[2]

Discussion

- WS indicates that the chemical could leach in soil, could run off, should not be adsorbed, might be biodegraded, and should not bioaccumulate. This chemical could contaminate the aquatic environment.

Exposure

Exposure could be by consumption of contaminated food and water if such contamination occurred. If the chemical were volatile (no volatility data are given here), then inhalation would be a problem.

Terbufos (*S*-[[(1,1-Dimethylethyl) thio] methyl] O,O-diethyl phosphorodithioate) (also Counter)

Data

Pesticide		
WS	12 ppm	[6]

Discussion

- This chemical could be adsorbed in soil, run off with soil, and be bioaccumulated, and it should not leach and should not be biodegraded. The chemical, if released into the environment, could contaminate the food chain. One piece of data is not enough for a conclusive prediction; additional data are needed.

Exposure

Routes of exposure could be by (1) consumption of contaminated water and food if such occurred, and (2) inhalation if the chemical were volatile; however, no data are presented herein on volatility.

Tetracene (also Naphthacene)

Data

WS	0.0005 ppm	[6]
Koc	650,000	[6]
Kow	800,000	[6]

Discussion

- WS indicates that this chemical should adsorb to soil, run off with soil, and be bioaccumulated, and it should not leach and should not be biodegraded. This prediction is supported by both Koc and Kow values.
- Koc indicates that the chemical should adsorb to soil.
- Kow indicates that the chemical should bioaccumulate and cause contamination of the food chain.
- The data support the prediction that this chemical should bioaccumulate in the food chain; thus residues could be expected in the food chain. This chemical should be persistent in the environment, and very little biodegradation and metabolism should be expected to reduce the residues.

Exposure

Routes of exposure could be by (1) consumption of contaminated water and food if such occurred, and (2) inhalation the chemical were if volatile; however, no data are presented herein on volatility.

1,2,4,5-Tetrachlorobenzene

Data

WS	6 ppm	[6]
Kow	47,000	[6]
BCF—flowing water	4,500	[6]

Discussion

- WS indicates that this chemical should adsorb to soil, run off with soil, and be bioaccumulated, and it should not be biodegraded and should not be leached. This prediction is supported by Kow and BCF values.
- Kow indicates that the chemical should bioaccumulate and could contaminate the food chain.
- BCF indicates that the chemical should bioaccumulate, and residues should be expected in the food chain.
- The data indicate that this chemical should bioaccumulate, cause contamination in the food chain, and have residues in the food chain if it were released to the environment.

Exposure

Routes of exposure could be by (1) consumption of contaminated water and food, and (2) inhalation if the chemical were volatile; however, no data are presented herein on volatility.

1,1,2,2-Tetrachloroethane

Data

WS	2,900 mg/l	[4]
Koc	400	[4]
VP	5 torr at 25°C	[4]
Kow	363	[4]

Discussion

- WS indicates that leaching, runoff, potential for biodegradation, and soil adsorption should not occur.
- Koc indicates that adsorption should not occur.
- Kow indicates that bioaccumulation should not occur.
- VP indicates volatility. This chemical is volatile and could fall out onto noncontaminated areas.

Exposure

The major route of exposure should be by inhalation. If the chemical were phototransformed to a nontoxic substance, inhalation might not be a problem unless an immediate one. Contamination of noncontaminated areas by fallout is possible, and this may another route of exposure.

Tetrachloroethylene (also Perchloroethylene)

Data

WS	150 to 200 mg/L	[4]
Kow	758	[4]
VP at 20°C	14 torr	[4]

Discussion

- WS indicates that this chemical could go either way in regard to leaching, runoff, adsorption, biodegradation, and bioaccumulation.
- Kow indicates that bioaccumulation could occur.
- VP indicates that this chemical is volatile, and it could present inhalation problems. If it were phototransformed, fallout of photoproducts and the parent chemical could contaminate water and food.
- This chemical is volatile and could present inhalation problems with the parent and potential photoproducts. Although it may bioaccumulate, it is unlikely to do so because it should volatilize prior to bioaccumulation. The volatility should also prevent residues from occurring in food, but there could be some in water if aquatic contamination occurred.

Exposure

Routes of exposure could be by (1) inhalation of volatile residues and (2) consumption of contaminated water and food if such occurred.

Thiabendazole

Data

WS	<50 ppm	[6]
Koc	1,720	[6]

Discussion

- WS indicates that this chemical could go either way in regard to leaching, runoff, adsorption, biodegradation, and bioaccumulation.
- Koc indicates that the chemical should adsorb to soil.
- Data are lacking that would permit a conclusive prediction. However, based on the Koc value, it is predicted that this chemical should be adsorbed to soil, should be bioaccumulated, and may cause contamination in the food chain if released into the environment.

Exposure

Routes of exposure could be by (1) consumption of contaminated water and food if such occurred, but additional data are needed to discern such contamination; and (2) inhalation if the chemical were volatile, but no data are presented herein on volatility.

Tillam (also *S*-Propyl butylethylthiocarbamate and pebulate)

Data

Pesticide

WS	60 ppm	[6]
Koc	630	[6]

Discussion

- WS indicates that this chemical could go either way in regard to leaching, runoff, adsorption, biodegradation, and bioaccumulation.
- Koc indicates that the chemical could leach, run off, and be biodegraded, and it should not be adsorbed and should not be bioaccumulated.
- Based on the value for Koc, it is predicted that this chemical could be mobile and could contaminate the food chain. Biodegradation and perhaps metabolism could be rapid enough to prevent exposure to residues; however, no data are presented herein to indicate this. It is predicted that residues may occur in the food chain if they are not dissipated rapidly.

Exposure

Routes of exposure could be by (1) consumption of contaminated water and food if such occurred, and (2) inhalation if the chemical were volatile; however, no data are presented herein on volatility.

Toluene (also Toluol)

Data

WS at 25°C	534.8 mg/L	[4]
Kow	490	[4]
VP at 25°C	28.7 torr	[4]

Discussion

- WS, supported by Kow, indicates that this chemical could leach, run off, and be biodegraded, and it should not adsorb in soil and should not be bioaccumulated.
- Kow indicates that bioaccumulation should not occur; however, residues could be expected in the food chain.
- VP indicates volatility; fallout could contaminate food and water. If the chemical were phototransformed, transformation products also could fall out and contaminate food and water.

Inhalation of the parent chemical and potential photoproducts could be a problem.

Exposure

Routes of exposure could be by (1) ingestion of contaminated food and water if such occurred, and (2) inhalation of volatile residues.

Toxaphene (Technical chlorinated camphene (67–69% chlorine))

Data

Pesticide

WS	0.4 ppm	[6]
BCF—flowing water	26,400	[6]
BCF—static water	4,250	[6]

Discussion

- WS indicates that this chemical could adsorb in soil, run off with soil, and be bioaccumulated, and it should not leach and should not be biodegraded.
- BCF indicates that bioaccumulation and food-chain contamination could occur. Residues could be expected in the food chain.

Exposure

Routes of exposure could be by (1) ingestion of contaminated food and water if such occurred, and (2) inhalation if the chemical were volatile; however, no data are presented herein to indicate volatility.

Triallate

Data

WS	4 ppm	[6]
Koc	2,220	[6]

Discussion

- WS indicates that this chemical should adsorb to soil, runoff with soil, and be bioaccumulated, and it should not be biodegraded and should not be leached.
- Koc indicates that the chemical should adsorb to soil, and this prediction is in agreement with the value for WS.
- This chemical could bioaccumulate and cause contamination of the food-chain; however, additional data are needed for a conclusive prediction. If such contamination occurred, then residues could be expected in the food chain.

Exposure

Routes of exposure could be by (1) consumption of contaminated water and food if such contamination occurred, but additional data are needed; and (2) inhalation if the chemical were volatile, but no data are presented herein on volatility.

1,2,4-Trichlorobenzene

Data

WS	30 ppm	[6]
Kow	15,000	[6]
Kow	18,197	[4]
BCF—flowing water	491	[6]
VP at 20°C	0.42 torr	[4]

Discussion

- WS indicates that this chemical could go either way in regard to leaching, runoff, adsorption, biodegradation, and bioaccumulation.
- Kow indicates that the chemical should bioaccumulate.
- BCF indicates that the chemical should have residues in the food chain but not bioaccumulate.
- VP indicates volatility; there is the potential for an inhalation problem. If the chemical were phototransformed, the phototransformates and the parent chemical could fall out onto water and food and thus contaminate the food chain.

- The data indicate that this chemical should have residues in the food chain if the chemical were released into the environment. Bioaccumulation may be prevented by metabolism in animals; however, no data are presented so that one can discern this. Volatility could be an inhalation problem, and fallout of residues could cause food-chain contamination.

Exposure

Routes of exposure could be by (1) consumption of contaminated water and food and (2) inhalation; however, additional data are needed to permit assessment of the potential problems associated with volatility.

Trichlorofluoromethane (also Freon-11 and Fluorocarbon-11)

Data

WS	1,100 mg/L	[4]
Kow	338.84	[4]
VP at 20°C	667.4 torr	[4]

Discussion

- WS indicates that this chemical should leach, run off, and be biodegraded, and it should not be bioaccumulated and should not be adsorbed in soil.
- Kow indicates that this chemical should not bioaccumulate.
- VP indicates volatility, and this could cause inhalation problems. If it were phototransformed, fallout of photoproducts and the parent chemical could contaminate water and food.
- This chemical is volatile, and the parent chemical and potential photoproducts could cause inhalation problems. The major problem should be degradation of the ozone layer in the atmosphere. Residues that fall out onto water and food should volatilize; however, it may take longer for volatilization to occur in water than it would take on food.

Exposure

Route of exposure should be by inhalation of volatile residues. The major concern should be degradation of the ozone layer.

Trichlorofon (also Dimethyl (2,2,2-Trichloro-1-hydroxyethyl) phosphonate, Dylox, Chlorphos, and Anthon)

Data

WS	154,000 ppm	[6]
Kow	3	[6]

Discussion

- WS indicates that this chemical should leach, run off, and be biodegraded, and it should not bioaccumulate and should not be adsorbed to soil.
- Kow indicates that the chemical should not bioaccumulate, and this prediction is supported by the WS value.
- This chemical could contaminate the environment if released into it. Biodegradation, if rapid, should prevent residues from being found in the food chain; however, no data are presented herein to support this prediction.

Exposure

Routes of exposure could be by (1) consumption of contaminated water and food if such occurred, and (2) inhalation if the chemical were volatile; however, no data are presented herein on volatility.

2,4,6-Trichlorophenol

Data

WS at 25°C	800 mg/L	[4]
Kow	2,399	[4]
VP at 76.5°C	1 torr	[4]

Discussion

- WS, supported by Kow, indicates that this chemical could adsorb in soil, run off with soil, and be bioaccumulated, and it should not leach and should not be biodegraded.
- Kow indicates that bioaccumulation and food-chain contamination could occur. Residues could be expected in food and water.
- VP indicates volatility; fallout could contaminate food and water. If the chemical were phototransformed, transformation products could also fall out and contaminate food and water. Inhalation of the parent chemical and its potential photoproducts could be a problem.

Exposure

Routes of exposure could be by (1) ingestion of contaminated food and water if such occurred; also, (2) inhalation could be a problem.

Triclopyr ([(3,5,6-Trichloro-2-pyridinyl) oxy] acetic acid) (also Garlon)

Data

Pesticide		
WS	430 ppm	[6]
Koc	27	[6]
Kow	3	[6]
BCF—static water	0.02	[6]

Discussion

- WS indicates that this chemical could go either way in regard to leaching, runoff, adsorption, biodegradation, and bioaccumulation.
- Koc indicates that the chemical should not adsorb and should not be bioaccumulated; however, it could leach, run off, and be biodegraded.
- Kow indicates that the chemical should not bioaccumulate.
- BCF indicates that it should not bioaccumulate; however, residues could be expected in the food chain.

- Residues of this chemical could be expected in the food chain because the chemical should be mobile. These residues could be short-lived if the chemical were biodegraded and metabolized in animals. The BCF indicates metabolism in animals.

Exposure

Routes of exposure could be by (1) consumption of contaminated water and food if such occurred, and (2) inhalation if the chemical were volatile; however, no data are presented herein on volatility.

Triclopyr (butoxyethyl ester)

Data

Pesticide		
WS	23 ppm	[6]
Kow	12,300	[6]

Discussion

- WS indicates that this chemical could go either way in regard to leaching, run off, adsorption, biodegradation, and bioaccumulation.
- Kow indicates that this chemical should bioaccumulate and should not be biodegraded, and perhaps should not be metabolized in animals.
- This chemical, if released into the environment, should bioaccumulate and could contaminate the food chain. Residues could be expected in the food chain.

Exposure

Routes of exposure could be by (1) consumption of contaminated water and food and (2) inhalation if the chemical were volatile; however, no data are presented herein on volatility.

Triclopyr (triethylamine salt)

Data

Pesticide		
WS	2,100,000 ppm	[6]
Kow	3	[6]

Discussion

- WS indicates that this chemical should leach, run off, and be biodegraded, and it should not adsorb in soil and should not be bioaccumulated.
- Kow indicates that the chemical should not bioaccumulate.
- This chemical should not be bioaccumulated in the food chain; however, this does not preclude residues in the food chain. The chemical should be biodegraded and may be metabolized in animals. Additional data are needed to support this prediction. This chemical should be mobile in the environment unless biodegradation were rapid.

Exposure

Routes of exposure could be by (1) consumption of contaminated water and food if such occurred, and (2) inhalation if the chemical were volatile; however, no data are presented herein on volatility.

Trietazine (2-chloro-4-diethylamino-6-ethylamino-*s*-triazine)

Data

Pesticide		
WS	20 ppm	[6]
Koc	600	[6]
Kow	2,200	[6]
Soil TLC-Rf	0.36	[1]

Discussion

- WS, supported by Kow and soil TLC, indicates that this chemical could bioaccumulate and run off with soil, and it should not leach and should not be biodegraded.
- Koc indicates that there should be no adsorption to soil; however, soil TLC and WS indicate that some may occur. Because two tests indicate that some adsorption should occur, the prediction should be that adsorption should occur to some extent. Remember that a negligible amount of chemical usually can go either way.
- Kow indicates that bioaccumulation and food-chain contamination could occur. Residues could be expected in the food chain.
- Soil TLC that indicates slight movement and adsorption could occur.
- There are contradictions between the above data; however, the data indicate that bioaccumulation and food-chain contamination could occur. Residues could be expected in food and water. Some adsorption to soil and some leaching could occur.

Exposure

Routes of exposure could be by (1) ingestion of contaminated food and water if such occurred, and 2) inhalation if the chemical were volatile; however, no date are presented herein to indicate volatility.

Trifluralin (α,α,α-trifluoro-2,6-dinitro-N,N-dipropyl-*p*-toluidine)

Data

Pesticide		
WS	1×10^{-4} g/100 ml at 27°C	[2]
VP	199 mm Hg $\times\ 10^{6}$ at 29.5°C	[2]
Degraded by microbes		[2]
Phototransformed		[2]
Adsorbs on organic matter		[2]
Soil TLC-Rf	0; class 1	[5]

Discussion

- WS indicates that there should be no leaching, and runoff should occur with soil particles, as supported by adsorption and soil TLC.
- Volatility could be a problem for inhalation and fallout onto noncontaminated areas.
- Phototransformation can occur, resulting in the formation of phototransformation products.
- Adsorption to soil has occurred.
- Soil TLC indicates that leaching should not occur.
- Based on the information presented, this chemical could bioaccumulate, volatilize, and cause food-chain contamination.

Exposure

Routes of exposure could be by (1) inhalation and (2) consumption of contaminated plants and animals. If fallout of volatile residues onto noncontaminated areas occurred, then food-chain accumulation would be likely.

Urea

Data

Pesticide		
WS	1,000,000 ppm	[6]
Koc	14	[6]
Kow	0.001	[6]

Discussion

- WS indicates that the chemical could leach and run off (as supported by Koc), may be biodegraded, and should not bioaccumulate (as supported by Kow).
- Koc indicates that adsorption should not occur.
- Kow indicates that bioaccumulation should not occur.
- This chemical could contaminate the aquatic environment.

Exposure

Exposure could be by (1) consumption of contaminated water if such contamination occurred. If the chemical were volatile (no volatility data are given here), (2) inhalation could be a problem.

Vinyl chloride (also Chloroethene)

Data

WS at 25°C	1.1 ppm	[4]
WS at 10°C	60 ppm	[4]
VP at 25°C	2,660 torr	[4]
Adsorbed light	at <220 nm	[4]

Discussion

- WS indicates that this chemical may or may not be mobile and may or may not bioaccumulate. Solubility appears to be a function of temperature, but this may be an incorrect assumption, depending on the controls used in testing. If the solubility was 1.1 ppm at 25°C, there might have been a loss due to VP. It was reported that at a lower temperature, 10°C, the solubility was 60 ppm, and the lower temperature could indicate a less volatile condition.
- VP indicates that the chemical could volatilize. Inhalation and fallout could be a problem.
- Phototransformation could occur in the troposphere above the ozone layer.
- Based on the data given volatilization, leaching, and bioaccumulation could be a problem.

Exposure

Routes of exposure could be by (1) inhalation and (2) consumption of contaminated plants, animals, and water if such contamination occurred. The volatile residue could fall out and contaminate noncontaminated areas.

Warfarin (3-(α-Acetonylbenzyl)-4-hydroxycoumarin)

Data

Pesticide—rodenticide	
Drug—anticoagulant	[8]
Freely soluble in alkaline solutions	[8]
Almost insoluble in water	[8]

Discussion

- Warfarin is almost insoluble in water; thus it should not leach, and runoff should be with soil particles. It should adsorb to soil organic matter and bioaccumulate.
- Under alkaline conditions it is freely soluble in water; thus leaching and runoff are likely, but adsorption and bioaccumulation are unlikely.

Exposure

Exposure could be by consumption of contaminated water, plants, and animals. Inhalation could be a problem; however, no data are presented herein on volatility.

REFERENCES

[1] Dragun, James and Charles S. Helling, "Evaluation of Molecular Modelling Techniques to Estimate the Mobility of Organic Chemicals in Soils: II Water Solubility and the Molecular Fragment Mobility Coefficient," Procedures, 7th Annual Residue Symposium, Land Disposal of Municipal Solid and Hazardous Waste and Resource Recovery, (unpublished paper), March 1981.

[2] Environmental Protection Agency, "Adsorption Movement and Biological Degradation of Large Concentrations of Selected Pesticides in Soils," EPA-600/2-80-124, August 1980.

[3] Environmental Protection Agency, "Sorption Properties of Sediments and Energy-Related Pollutants," EPA-600/3-80-041, April 1980.

[4] Environmental Protection Agency, "Water-Related Environmental Fate of 129 Priority Pollutants," Contract No. 68-01-3852, Draft, January 1979.

[5] Helling, Charles S. and Benjamin C. Turner, "Pesticide Mobility: Determination by Soil Thin-Layer Chromatography," *Science,* vol. 162, pp. 562–563, November 1968.

[6] Kenaga, E. E. and C. A. I. Goring, "Relationship between Water Solubility, Soil Sorption, Octanol–Water Partitioning, and Concentration of Chemicals in Biota," Special Technical Publication 707, American Society for Testing Materials, 1980.

[7] Lyman, Warren J., William F. Reehl, and David H. Rosenblatt, *Handbook of Chemical Property Estimation Methods,* McGraw Hill Book Company, 1982.

[8] *The Merck Index,* Ninth Edition, Merck & Company, 1976.

[9] Ney, Ronald E., Jr., "Fate, Transport and Prediction Mode Application to Environmental Pollutants," Spring Research Symposium, James Madison University, UP, April 16, 1981.

[10] Yip, George and Ronald E. Ney, Jr., "Analysis of 2,4-D Residues in Milk and Forage," *Weeds, Journal of the Weed Society of America,* vol. 14, pp. 167–170, 1966.

Appendix
List of Chemicals Included in Chapter 5

Acenaphthene
Acenaphthylene
Acephate
Acetophenone
Acridine
Acrylonitrile
Alachlor
Aldicarb
Aldrin
Amitrole
Ammonia
Aniline
Anthracene
Asbestos
Atrazine
Benefin
Bentazon
Benzene
Benzidine
Benzo [a] anthracene
Benzo [b] fluoranthene
Benzo [a] pyrene
Bifenox
Biphenyl
Bromacil
Bromobenzene
Bromodichloromethane
Bromoform
Bromomethane
4-Bromophenyl phenyl ether
Bufencarb
Butralin
Butyl benzyl phthalate
x-*sec*-Butyl-4-chlorodiphenyloxide
Captan
Carbaryl
Carbofuran
Carbon tetrachloride
Chloramben
Chlorbromuron
Chlordane
Chlorine
Chlorobenzene
4-Chlorobiphenyl
p-Chloro-*m*-cresol
4-Chlorodiphenyl oxide
Chloroethane
Bis (2-Chloroethoxy) methane
Bis (2-Chloroethyl) ether
Chloroform
Chloromethane
Bis (2-Chloromethyl) ether
Chloroneb
2-Chlorophenol
4-Chlorophenyl phenyl ether
Chloroxuron
Chlorpropham
Chlorthiamid
Chrysene
Crufomate
Cyanazine
Cycloate
2,4-D
Dalapon
DBCP
DDD
DDE
DDT
Dialifor
Diallate
Diamidofos
Diazinon
Dibromochloro methane
Dicamba
Dichlobenil
Dichlofenthion
o-Dichlorobenzene
p-Dichlorobenzene
4,4′-Dichlorobiphenyl
Dichlorodifluoro methane

2,4-Dichlorophenol
3,6-Dichloropicolinic acid
Dichlorovos
Dieldrin
Diethylanaline
Di-2-ethylhexyl phthalate
Diethyl phthalate
Dimethoate
Dimethylnitrosoamine
Dimethyl phthalate
Dimilin
Dinitramine
4,6-Dinitro-*o*-cresol
2,4-Dinitrophenol
2,4-Dinitrotoluene
Dinoseb
Diphenylnitrosamine
Diphenyl oxide
Di-*n*-propylnitrosamine
Disulfoton
Diuron
DSMA
Dursban
Endothall
Endrin
EPTC
Ethion
Ethylbenzene
Ethylene dibromide
Ethylene dichloride
Fenuron
Fluchloralin
Fluoranthene
Fluorene
Formetanate
Glyphosate
Heptachlor
Hexachlorobenzene
Hexachlorobutadiene
Hexachlorocyclo pentadiene
Hexachloroethane
Imidan
Ipazine
Isocil
Isopropalin
Kepone
Lead
Leptophos
Lindane
Linuron
Malathion
Methazole
Methomyl
Methoxychlor
2-Methoxy-3,5,6-trichloropyridine
9-Methylanthracene
Methyl chloroform
3-Methylcholanthrene
Methylene chloride
Methyl isothiocyanate
2-Methylnaphthalene
Methylparathion
Metobromuron
Metolachlor
Mexacarbate
Mirex
Monolinuron
Monuron
Naphthalene
1-Naphthol
Nitralin
Nitrapyrin
Nitrobenzene
2-Nitrophenol
4-Nitrophenol
Norflurazon
Oxadiazon
Paraquat
Parathion
Pentachlorobenzene
Pentachlorophenol
Phenanthrene
Phenol
Phosalone
Phthalic anhydride
Phorate
Picloram
Polychlorinated biphenyls (PCB's)
Profluralin
Propham
Propoxur
Pyrazon
Pyrene
Pyroxychlor
Radon
Ronnel
Silvex
Simazine
2,4,5-T
TCE
Terbicil
Terbufos
Tetracene
1,2,4,5-Tetrachlorobenzene
1,1,2,2-Tetrachloroethane
Tetrachloroethylene and Perchloroethylene
Thiabendazole
Tillam
Toluene
Toxaphene
Triallate
1,2,4-Trichlorobenzene
Trichlorofluoromethane

Trichlorofon
2,4,6-Trichloro-
  phenol
Triclopyr
Triclopyr
  (butoxyethyl ester)
Triclopyr
  (triethylamine salt)
Trietazine
Trifluralin
Urea
Vinyl chloride
Warfarin

Glossary

Absorption. The process by which chemicals are temporarily held in soil, plants, or animals, but can be released.

Adsorption. The condition of being bound by a particular mechanism and not released with water (*see* Bound).

Accumulation. The buildup or increase of a chemical(s) or its breakdown product(s) in soil or particulate matter.

Attenuation. The physical, chemical, and biological functions of soil(s) that prevent chemical movement up, sideways, or downward through the soils(s).

Biodegradation. The process by which chemicals are degraded by microbiological organisms (aerobic and/or anaerobic microbes) to produce a lower-molecular-weight chemical(s) called a biodegradate(s).

Bound. The process by which a chemical is adsorbed to soil (e.g., to organic carbon, by ion exchange, etc.) or adsorbed in plants or animals (e.g., in protein, etc.).

Breakdown products. Products of the breaking down of a parent chemical to a lower-molecular-weight chemical(s). These products are called degradates, biodegradates, metabolites, transformation products, hydrolytic products, dissociation products, etc.

Degradation. Chemical or physical process by which chemicals are altered to a lower-molecular-weight chemical(s) called a degradate(s).

Detoxification. The altering of a toxic chemical by degradation, microbiological degradation, or biodegradation, metabolism, photolysis, hydrolysis, dissociation, etc., to a nontoxic chemical.

Dissipation. The disappearance of a chemical and/or its breakdown products from environmental compartments.

Environment. Air, surface water (lakes, streams, ponds, etc.), groundwater, soil, plants, and animals, including humans.

Environmental compartments. Air, water, soil, plants, and animals.

Half-life. The time it takes for a chemical and/or its breakdown product(s) to dissipate to 50% of its original amount.

Leaching. The movement of a chemical through some medium (e.g., soil) to a noncontaminated area.

Lipo-soluble. Fat-soluble.
Metabolism. The process by which plants or animals change a chemical(s) to a higher- or lower-molecular-weight chemical(s).
mg/L. Milligrams per liter.
Mobility. The movement of a chemical or breakdown product(s) by runoff, volatilization, and leaching.
Parent chemical. The chemical from which a breakdown product is formed; the starting chemical.
Photolysis. The physical reaction that occurs when a chemical sorbs light waves, causing it to transform to a higher- or lower-molecular-weight chemical(s).
ppb. Parts per billion.
ppm. Parts per million.
Radiolabeled chemical. A chemical that is intentionally made radioactive.
Residue. Remnant; in this context, a chemical found in an environmental compartment.
Runoff. The movement of a chemical over some medium (e.g., soil surface, plant surface) to a noncontaminated area.
Sorbed. Taken up and held by either adsorption or absorption.
Synergistic effects. Interaction of two or more substances.
Transformation. The process by which a chemical(s) is altered to form a higher- or lower-molecular-weight chemical(s) called a transformation product(s).
Translocation. The movement of a chemical within a plant or an animal.
μg/L. Micrograms per liter.
Vapor pressure. A measure of volatility expressed in mm Hg (millimeters of mercury) or torr. One torr is the same as one mm Hg at zero degrees Celsius and at sea level.
Volatility. The ability of a chemical(s) to change to a gas and move to a noncontaminated environment.
Water. An all-inclusive term herein, used to include all forms of water (streams, ponds, groundwater, oceans, etc.).

Index

About the Author

The author, Ronald Ellroy Ney, Jr., Ph.D., worked for the federal government for twenty-five years before retiring from government service in September 1986. Currently, Dr. Ney is a real estate broker, certified real estate appraiser, consultant on chemicals in the environment, and a registered environmental assessor (State of California).

His work experience has included service with:

- The U.S. Environmental Protection Agency
  - He was a technical advisor in the Office of Solid Waste, Land Disposal Branch in the development of hazardous waste research needs and policies as related to fate and transport chemistry, modeling, monitoring, and hazard-risk assessment. He served as liaison person for Center of Excellence Research Centers, the Dioxin Disposal Advisory Group, Pesticide Disposal Workshops, and the Pesticide in Ground Water Work Group.
  - He was a section chief in the Hazard Evaluation Division, Environmental Fate Branch for the evaluation of scientific data. He wrote methodology protocols to obtain data on the fate and transport of organic and inorganic pesticides, in air, water, soils, plants, and animals and predictive models.
  - He was a section chief in the Pesticide Registration Division, Efficacy and Ecological Effects Branch for the evaluation of scientific data. He wrote methodology and guideline protocols to obtain data on fate and transport to support pesticide registration.
- The U.S. Department of Agriculture
  - He served as supervisory chemist in the Pesticide Registration Division, Chemistry Staff and was responsible for pesticide

petitions for tolerance review and label review. It was here that Dr. Ney developed the first data requirements and founded the first environmental chemistry program on fate and transport of chemicals in the environment.

- U.S. Department of Welfare, Food and Drug Administration
  - He served as a laboratory group leader and chemist on pesticide residue methodology
- Fairfax County, Virginia School System and Harrisonburg, Virginia School System
  - He taught science and chemistry.

Dr. Ney was included in the Sixteenth Edition of the *American Men and Women of Science,* January 1987, for outstanding contributions in physical and biological sciences.